Welding of plastics

Core research from TWI

A selection of reports presenting research findings from highly professional and skilled engineers, originally prepared under TWI's core research programme for industrial members.

ABINGTON PUBLISHING
Woodhead Publishing Ltd in association with TWI Ltd
Cambridge England

Published by Abington Publishing
Woodhead Publishing Limited, Abington Hall,
Abington, Cambridge CB1 6AH, England
www.woodhead-publishing.com

First published 2000, Abington Publishing

British Library Cataloguing in Publication Data
A catalogue record for this book is available from the British Library.

ISBN 1 85573 519 9

Printed by Victoire Press, Cambridge, England.

CONTENTS

STRUCTURAL INTEGRITY OF BUTT FUSION WELDED POLYETHYLENE PIPES - A REVIEW

By: M J Troughton and G S Booth

TECHNOLOGY BRIEFING

A review of the research which has been undertaken, both in industry and academia, regarding the structural integrity of butt fusion welds in polyethylene pressure pipes has revealed that current tests are inadequate for assessing the service performance of such joints.

Background

In many industrial situations, plastics are being used to replace traditional materials such as metals, both for economic reasons and for their improved performance and processing characteristics. They are being applied increasingly for more demanding applications such as the use of polyethylene (PE) in pressurised pipe systems for the transportation of water, gas and chemicals. Such developments place a heavy requirement on the provision of adequate design data to give the user confidence in the safety of the system.

The life of a plastics pipe system depends upon a number of factors, the most important being the properties of the pipe material and joints, flaws and stresses in the system and the service temperature. Ideally, design data should take into account these factors when assessing the durability of the system.

A great deal of work has been performed over the past 20 years to assess the durability of the pipes themselves. Considerably less work has been performed on the welded joints.

Objective

- To review the current situation regarding methods for assessing the structural integrity of butt fusion welds in PE pipes.

Approach

The structural integrity of a butt fusion weld in a PE pipe depends on three factors: the existence of any defects, the service conditions and the properties of the joint.

The main sources of defects in PE pipe butt fusion welds are described, together with the currently available non-destructive methods for detecting them. This is followed by an explanation of the various possible stresses which may be present in a welded PE pipe system in service. Finally the test methods used for assessing the structural integrity of butt fusion welds in PE pipes are discussed.

Discussion

To predict the service life of butt fusion welds in PE pipes requires that long-term tests are performed under conditions which allow reliable extrapolation of results back to those conditions which occur in service.

Long-term tests on coupons cut from the weld suffer from the fact that the stresses in a coupon may well be different to those in the whole welded pipe.

Long-term hydrostatic pressure tests on whole butt fusion welded pipe sections are also inadequate for predicting the service life of the joint since, invariably, failure occurs in the pipe wall first. Failure at the weld can be generated using hydrostatic pressure tests on pipes with a radial offset at the joint. However, the misalignment produces an artificially sharp notch between the weld bead and the pipe wall which may not allow reliable extrapolation of data back to that for perfectly aligned joints.

Therefore, at present there is no test available which has been shown conclusively to predict the service life of weld-made butt fusion welds in PE pipes or, conversely, to estimate the maximum allowable flaw size for a specified lifetime and loading. Consequently, it is not known whether present NDT techniques are sufficiently sensitive to achieve on-site evaluation of pipe joints.

There are a number of short-term coupon tests commonly used for QA purposes and to optimise the butt fusion welding process. However, different tests give different answers for the 'strength' of a weld. For example, a good quality weld will achieve a welding factor of around 1.0 in most tensile tests, whereas the same weld would only achieve a 'welding factor' of around 0.5 in impact tests. Thus, there is a need to determine which of these short-term tests are the most useful.

Conclusions

- At present there are no reliable tests or procedures to determine the structural integrity of welded PE pipe systems.
- Since critical defect sizes are, as yet, unknown for butt fusion welds in PE pipes, it is not known whether current non-destructive testing techniques are sufficiently sensitive to detect them.
- Current tests on whole pipe sections do not adequately simulate the stresses which may cause premature failure of the butt fusion weld in service.
- There are a large number of tests for coupons cut from a pipe joint. However, the value of weld strength obtained depends on the test itself.

Recommendations

It is recommended that a large-scale test be developed which can reliably predict the service life of butt fusion welds in PE pipes. The results from this test should then be correlated with results from the many short-term coupon tests currently used in industry in order to determine which of the coupon tests gives the best indication of the long-term durability of the weld and should therefore be used for QA purposes and to optimise butt fusion welding parameters.

STRUCTURAL INTEGRITY OF BUTT FUSION WELDED POLYETHYLENE PIPES - A REVIEW

By: M J Troughton and G S Booth

1. INTRODUCTION

Plastics materials are being used in an increasing number of applications, although to date many of these applications do not involve structural integrity issues. One important exception, however, is the use of polyethylene (PE) pipe systems for the distribution of water and gas under internal pressure. Under these circumstances, an understanding of the long term performance of PE pipelines (and welds in particular) is vital to ensure premature failures do not occur.

The aim of this report is to review the progress which has been made regarding methods of determining the structural integrity of butt fusion welded PE pipe systems used for the distribution of water and gas. Types of flaws in butt fusion welds, service conditions and performance assessment methods are considered.

It is widely recognised that the welding process may introduce flaws, which may significantly influence joint strength. For metals, a methodology has been developed, based on fracture mechanics principles, for assessing the significance of a flaw on joint fracture and fatigue strength (1). However, such a methodology has not yet been fully developed for non-metallic materials.

1.1. Polyethylene: Structure and Applications

Polyethylene (PE) is produced by polymerizing ethylene gas in the presence of a catalyst under controlled temperatures and pressures (see Fig. 1). The structure of the molecule can be altered, depending on the catalyst and reaction conditions used. Relatively low pressures and temperatures, combined with a Ziegler–Natta catalyst, produce a linear molecule and a 'high density' polyethylene (HDPE); whereas higher pressures and temperatures and the presence of peroxide–type catalysts produce a relatively highly branched molecule and a polyethylene with a lower density (LDPE). Intermediate degrees of branching can be produced by adding small percentages of alpha–olefin comonomers (such as butene or hexene) to the ethylene. The resulting materials, such as 'medium density' polyethylene (MDPE), which also includes linear low density polyethylene (LLDPE), have intermediate densities (see Fig. 2).

The reason why the degree of molecular branching affects the density of the material is that PE is a semi–crystalline polymer, which means that it contains areas where the molecular chains are well–ordered and packed tightly together; the crystalline regions. However, not all of the molecules can attain close molecular alignment due to the presence of side branches off the main molecular 'backbone', resulting in unordered, amorphous regions. Where the molecules are packed more tightly, i.e. in the crystalline regions, the density is higher. Therefore, the higher the crystallinity, i.e. the percentage of material which is in the crystalline regions, the higher the density.

As the density of PE increases so too does its tensile strength, rigidity, abrasion resistance, barrier properties and heat resistance. Decreasing the density leads to increased toughness, flexibility and resistance to environmental stress cracking.

Such a range of properties, which also includes:-

- low cost
- excellent processability
- excellent electrical insulation properties
- excellent chemical resistance
- freedom from odour and toxicity

has made PE the most widely used plastics material in the world today, with applications in almost every industry. Some of the major applications are given in Table 1.

1.2. Polyethylene Pipes

1.2.1. History

The first PE pipes were introduced in the mid-1940s (see Table 2). These were of small diameter and made from LDPE. HDPE pipe systems have been available since the 1950s although it is only over the last 20 years, mainly due to the development of MDPE pipe grades for gas distribution pipelines, that the use of PE has seen a dramatic increase (in Europe between 1973 and 1984, consumption of MDPE and HDPE for pipe applications rose from about 30,000 to 132,000 tonnes). In 1982/3 MDPE pipes were also introduced by the UK Water Companies. More recently, in the late 1980's, a new 'higher performance' HDPE was introduced for gas and water pipes.

1.2.2. Properties

The main features which make PE an attractive material for water and gas distribution pipes include (3):-

- flexibility
- light weight - easier to handle and transport
- good joining methods available
- good low temperature impact properties
- excellent chemical resistance
- good corrosion resistance
- relative impermeability to the fluids carried
- ability to withstand both internal and external loads

- non–toxicity

However, unlike most metals under normal conditions, the behaviour of PE, as for all thermoplastics, is highly time and temperature dependent. The result of this is that a PE pipe under internal pressure will fail after a certain time. Therefore, in contrast to metallic pipes, the lifetime of a PE pipeline is mainly determined by its long–term response to the applied stresses.

1.2.3. Pipe design

The relationship between the internal pressure, P, in a thin–walled pipe, and the stresses in the pipe wall (see Fig. 3) is given by the following equations:–

$$\sigma_h = \frac{P\,(D-t)}{2t} \qquad [1]$$

$$\sigma_a = \frac{P\,(D-t)}{4t} = \frac{\sigma_h}{2} \qquad [2]$$

where
σ_h = hoop stress (MPa)
σ_a = axial stress (MPa)
P = internal pressure (MPa)
D = external pipe diameter (mm)
t = minimum wall thickness (mm)

Hence, pipes under internal pressure are subjected to a biaxial stress state; with the hoop stress being twice the axial stress.

PE pipeline systems for gas and water applications are typically intended to endure 50 years at a temperature of 20°C. Considering the time dependent behaviour of plastics pipes, it is clearly necessary to determine their resistance to internal pressure during long term operation. The method currently used by the industry is to expose pipe samples to various internal hydrostatic pressures, in excess of those to be used in service, determine the hoop stress from equation [1], and measure the time it takes for the pipe to burst (4–6). The results from these tests are then extrapolated (7) to give a value for the minimum required strength (MRS) (8) of the pipe material after 50 years. This method will be covered in more detail in section 4.3. The present classification of pipe materials is given in Table 3.

The MRS can then be related to a hydrostatic design stress (HDS) (8) using the equation:–

$$HDS = \frac{MRS}{CD} \qquad [3]$$

where CD is the design coefficient, or safety factor. For water pipes a minimum design coefficient of either 1.25 (8) or 1.66 (10) is used, whereas for gas pipes, where safety is more critical, typical minimum values of CD are between 2 and 3 (9, 11). The final safety factor is decided by the designer of each individual pipe system and may well be greater than these minimum values, depending on the application, installation, joint quality, operating temperatures and the consequence of a failure (12, 13).

It is the HDS which is used when calculating pipe dimensions for a particular application. Equation [1] can be rewritten:–

$$\sigma_h = \frac{P}{2}\left(\frac{D}{t} - 1\right) \qquad [4]$$

and hence, for a fixed allowable hoop stress, there is a direct relationship between the internal pressure and the ratio of pipe external diameter and wall thickness, ie. as the pressure is increased the pipe wall thickness must also increase for the same pipe diameter. For PE pipes, the ratio $^D/_t$ has a number of fixed values which are common to both gas and water applications. These are called the standard dimension ratios (SDRs); the most common being 11, 17.6 and 26. Equation [4] can therefore be rewritten as:–

$$HDS = \frac{Pw}{2}(SDR - 1) \qquad [5]$$

where Pw is the maximum allowable working pressure. The relationship between Pw and SDR, for various values of HDS, is given in Table 4, although it must be remembered that the actual value of allowable working pressure for a particular pipe application may be considerably lower than these maximum values (due to a higher CD), especially for larger diameter pipes.

Plastics pipes have been produced in Europe with diameters of at least 1600 mm, although 800 mm is probably the largest currently produced in the UK. Standard pipe dimensions up to 500 mm diameter are given in Table 5.

1.3. Butt Fusion Welding of Polyethylene Pipes

It is normally impractical to extrude plastics pipe on site, so it is cut into lengths for transport from the factory to site and is then joined. One of the major advantages of PE pipe systems is that they can be joined using fusion welding. This gives a sealed system with the virtual elimination of rubber ring joints, which have been the main cause of leakage problems for other pipe systems, e.g. iron and PVC.

The principle of the fusion welding technique is to melt the two pipe surfaces sufficiently so that molecular segments can diffuse across the contact surface and form bonds (14, 15), which can give a joint of comparable strength to that of the parent components.

Butt fusion joining is the primary technique used for large (>250mm) diameter PE pipes, and the only one for diameters greater than 500mm, although it can be used for systems down to 63mm.

The principal sequence of the butt fusion welding cycle is shown in Fig. 4. Prior to this sequence the pipes are mounted in the clamps of the butt fusion equipment (see Fig. 5) and checked for alignment. The pipe ends are then planed using a special tool to ensure that they are flat and square.

The welding sequence begins when a flat hot plate, normally at a temperature of between 195°C and 230°C, is positioned between the two pipe ends and the pipes are pushed towards each other using hydraulic rams; the drag pressure in Fig. 4 being the hydraulic pressure required to overcome the frictional drag of the pipe. The pipe ends then come into contact with the hot plate and the pressure is increased to give good thermal contact. The pipe ends melt and the interface pressure, typically 0.15 MPa (16), forces the molten material outwards to form 'weld beads' at the outside and inside pipe surfaces; hence the term 'bead–up' stage. When the bead size has reached a set value (typically 2mm for small diameter pipes) the pressure is reduced to a very low (almost zero) constant pressure, sufficient only to maintain the pipe in contact with the hot plate. This allows the melt depth to increase without increasing the size of the weld beads. The 'heat soak' time is typically between 2 and 5 minutes, depending on the size of pipe. After the heat soak stage the pipe ends are pulled away from the hot plate, the hot plate is removed, and the two molten pipe ends are pushed back together.

The pressure applied during this 'fusion' stage is normally the same as the bead–up pressure. It causes further growth of the weld bead and is normally maintained until the weld is fully cooled.

Studies have shown that joint strength is largely tolerant of many butt fusion welding conditions (17–23); the dominant factor being the time for which the two halves of the joint are together at a temperature above the crystallisation temperature, T_c (18, 19, 22, 23). The plate removal stage is therefore critical since, at this point, the molten pipe ends are exposed to air and start cooling at an exponential rate. Consequently, it is very important that this stage is kept as short as possible. However, removal of the hot plate is often a manual operation and can take well over 10 seconds. For this reason a sufficiently large volume of material is melted during the heat soak stage, to ensure that the pipe ends remain above T_c during the removal of the hot plate, which then requires a long cooling period during the fusion stage. This makes butt fusion welding a time consuming process, especially for large pipes (>250mm diameter), where welding cycles can take over an hour.

Fortunately, fully automatic butt fusion equipment is now available which can reproducibly reduce the hot plate removal time to ≤ 3 seconds (24), suggesting that more rapid welding cycles can be used (22, 25).

Since the welding process parameters for pipes of diameter greater than 250mm are not as well defined as for smaller sizes, a number of modified butt fusion welding cycles, such as the 'dual pressure' system (26) (see Fig. 6), have been developed for the welding of large pipes; especially those made from PE100 material. These procedures increase the time the weld interface temperature is above T_c by reducing the pressure during the fusion stage after a set time and, therefore, reducing the amount of molten material displaced into the weld beads (18, 27).

2. DEFECTS AND DEFECT DETECTION IN BUTT FUSION WELDS IN POLYETHYLENE PIPES

2.1. Sources of Defects

2.1.1. Weld beads

The notch created by the weld bead rolling towards the pipe surface has been shown to be the site of crack initiation (28–30). Removal of the bead has been shown to relocate failure away from the welded joint in fatigue tests (30, 31).

2.1.2. Cold welds

If, during the butt fusion welding process, either the heater plate temperature is too low or the plate removal phase is too long, the surfaces of the pipe ends will crystallise before they come into contact and a 'cold' weld will result (18, 26, 32, 33). Such welds tend to fail in a brittle manner at the weld interface in internal pressure tests (29).

2.1.3. Contamination at the weld interface

The welding of PE pipes is frequently carried out in all weathers with no protection for equipment during the welding operation (23). There is therefore the possibility of contaminants such as dust, sand, soil, water, oil, etc being introduced into the weld interface (23, 26, 32, 33). Other possible contaminants are degraded plastic, caused by excessive heater plate temperature or heating time (26); and small particles of aluminium, caused by wear of the planing cutter used to prepare the pipe ends for joining (34).

The effect of contamination on weld behaviour has been investigated by a number of researchers (23, 35, 36) who have shown that, for the same applied stress, the minimum size of a flaw at the weld interface necessary to induce a brittle failure in a tensile test reduces as the pipe wall thickness increases. There may be a number of reasons for this behaviour (23):

1) Thicker walled sections are more likely to be under plane strain conditions, due to high local constraint (triaxiality), and are therefore less affected by the ductile processes that occur in the surface regions.

2) The welding parameters for thicker walled pipes may not be sufficiently optimised, resulting in a joint with inferior toughness properties compared to pipes with thinner wall sections, even in the absence of contamination (35, 36).

3) Any contamination at the weld interface will be moved by melt shear towards the weld bead during the fusion stage. However, as the pipe wall thickness increases a comparatively smaller proportion of the melt is forced into the beads and any contamination will not be totally removed from the weld interface.

In an attempt to prevent contamination, which may be present on the heater plate, from reaching the weld, it has been recommended that at least one 'dummy weld' is made prior to production welding (16).

2.1.4. Misalignment

Misalignment of pipe ends during the welding process can occur for a number of reasons, including internal and external diameters being at the extremes of the tolerance range, pipes clamped out of alignment during welding, ovality, and damaged pipe ends. This results in a sharpening of the notch between the weld bead and the pipe wall (see Fig. 7), together with an increase in stress due to local bending (see section 3.3).

2.2. Non–Destructive Testing

Non–destructive testing (NDT) of welded joints in PE pipelines is relatively undeveloped compared with NDT of welds in metals and, although the importance of NDT is recognised by manufacturers and users of PE pipes, there are no nationally accepted test procedures. In the past, there have been attempts to introduce NDT procedures for the routine quality control of PE welds (37). However, other than simple visual inspection (38, 39), these have all proved unsuccessful. This was because equipment for the specific purpose of testing PE welds was not available, the results from the tests that were used were difficult to interpret and, since the significance of defects was not clearly understood, the level of sensitivity required was not defined (33).

At present, a number of NDT methods are used to assess weld quality in PE pipes. The most common include visual inspection (38, 39), ultrasonics (40–43) and radiography (44–46), but others, such as spark testing and holographic interferometry (47–49) have also been investigated. However, none of these can, as yet, detect all possible flaws in PE pipe welds.

2.2.1. Visual inspection

A field judgment of the quality of a butt fusion joint is frequently based upon a visual inspection of the weld and its external bead. This is a very valuable method and can detect a number of welding faults, including pipe misalignment, insufficient melting and contamination. For example, a bead which is not completely uniform around the circumference, or is not symmetrical, indicates pipe misalignment.

British Gas specifies that all external weld beads should be removed and examined (11, 38). The bead width is checked against standard ranges, such as those given in Table 6. Any deviation from these dimensions indicates either insufficient, or excessive, melting during welding. The underside of the bead is examined for contamination and circumferential slits (16, 38, 39) (see Fig. 8). Slit defects, which can be confirmed by bending the bead back on itself, can indicate a poor weld due to contamination (39) or the production of a 'cold weld' (34). Cold welds have also been revealed by mechanical testing (49, 51) of the weld bead. However, a visually good weld does not guarantee good joint quality.

The internal weld beads are also sometimes removed and examined. This has the advantage that the flow capacity of the pipe is improved and the notch between the weld bead and the pipe bore is eliminated.

2.2.2. Ultrasonic testing

The principle of ultrasonic examination is based on propagating a high frequency sound wave through the pipe wall and measuring the intensity of the reflected wave. Any defects in the wall will alter the speed of the ultrasonic wave, resulting in a deviation of the signal. Although this technique works well for metals, a number of problems are encountered when examining PE. These include high attenuation of the ultrasound (33), which can vary significantly from one PE grade to another (52), and variations in material properties through the pipe wall which affect the wave velocity (33).

Despite these difficulties ultrasonic testing has been used to detect some flaws in PE pipe welds, such as cracks and voids (41–43, 53–55).

2.2.3. Radiographic testing

For the testing of PE welds, radiography has many advantages over ultrasonics, especially when used in conjunction with visual examination (33). It produces a good image of volumetric flaws (17, 55) and of weld beads on both internal and external surfaces of the pipe. It also provides a permanent record of the test and it is more easily interpreted than ultrasonics techniques.

Since PE becomes transparent at the standard X–ray voltages used for metals (50kV), considerably lower energy levels have to be employed (33, 52). An advantage of using these 'soft' X–rays (10kV) is that they are less of a health hazard and can be produced by portable battery operated X–ray equipment (33, 52). However, at these voltages there is little standardisation in the test equipment and procedures (52).

3. SERVICE CONDITIONS

PE pipes are subjected to a host of stresses in service; all of which will affect the lifetime of the system. Some of the important stresses, and their origins, are described below.

3.1. Internal Pressure

As stated in Section 1.2.3, hoop and axial stresses are generated in a pipe wall due to internal pressure and are given by equations [1] and [2]. However, these equations are only approximations for the particular case of thin–walled pipes (ie. where the ratio of internal diameter to wall thickness is greater than 20) and assume that the hoop stress is constant through the pipe wall. For thick walled pipes (SDR<18), this assumption is no longer valid and the hoop stress varies through the pipe wall according to the equation:–

$$(\sigma_h)_r = P\left(\frac{r_i^2}{r_o^2 - r_i^2}\right)\left(1 + \frac{r_o^2}{r^2}\right) \qquad [6]$$

where
$(\sigma_h)_r$ = hoop stress at radius r
r_i = inside radius of pipe
r_o = outside radius of pipe

3.2. Residual Stress

PE pipes are manufactured by extruding molten plastic through an annular die. The pipe is subsequently externally cooled using either water baths and/or sprays. This causes the external surface layer of the pipe to solidify and contract, while simultaneously, the still molten inside surface layer compresses plastically. When the inside layer subsequently cools and solidifies it attempts to contract, but is prevented from doing so by the external layer. This generates a tensile stress on the inside surface of the pipe and a compressive stress on the outside surface (see Fig. 9). The magnitude and distribution of the residual stress are functions of wall thickness and cooling rate.

In order to prevent tensile residual stresses occurring at the inside wall the extruded pipe can be simultaneously cooled on the outside and inside (11, 56). This results in a residual stress distribution as shown in Fig. 10.

In addition, since the welding process involves differential cooling of constrained molten material, residual stresses will also be present in pipe welds.

Although little, if any, work has been performed to assess the nature or magnitude of residual stresses in pipe welds, for the pipe itself values have been determined using either a 'split–ring' test (10, 57) or by layer removal techniques (56, 58).

Residual stresses, however, are not constant in PE; relaxation occurs as a result of time dependent behaviour and residual stress levels measured 24 hours after extrusion will decay by a factor of two after approximately 10 years (2).

3.3. Misalignment

The effect of misalignment, in addition to sharpening the notch between the weld bead and the pipe wall (see section 2.1.4), is to produce secondary bending due to the offset of the pipe centrelines (see Fig. 11). This increases the axial stress at the weld (59, 60) according to the equation:–

$$\sigma_{aj} = \sigma_a(1 + 3^e/t) \qquad [7]$$

where
σ_{aj} = axial stress at welded joint
σ_a = axial stress in the pipe
e = maximum radial displacement
t = pipe wall thickness

Radial displacements of up to 20% of the pipe wall thickness have been observed in service (31), corresponding to a stress concentration factor (σ_{aj}/σ_a) of 1.6.

3.4. Bending

One of the main advantages of PE pipes is their flexibility. This means that they can be bent in order to change the direction of the pipeline or to conform to an uneven ditch bottom. Also, unintentional pipe bending may arise due to ground movement after installation. These factors introduce a tensile axial stress in the pipe wall on the outside of the bend and a compressive axial stress in the wall on the inside. Provided bending is small, the relationship between the radius of curvature, R, and the axial stresses in the pipe wall, σ_m, (see Fig. 12) can be given by the following equation (2):–

$$R = \frac{ED}{2\sigma_m} \qquad [8]$$

where E is a modulus and D is the external diameter of the pipe. As with residual stresses, these bending stresses relax with time.

Equation [8] can be rewritten in terms of strain, i.e.

$$R = \frac{D}{2\epsilon_m} \qquad [9]$$

where ε_m is the tensile strain in the outer pipe wall. The minimum allowed value of R for a cold bend in PE pipes has been specified as 10–15 times the pipe outside diameter (2, 50), which relates via equation [9] to a tensile strain of 3–5%.

3.5. Thermal Contraction

PE pipes installed in warm weather develop a tensile axial stress when buried, as the pipe attempts to contract at the lower underground temperature but is prevented from doing so by the surrounding soil. This effect is more pronounced in PE pipes than in metallic pipes due to their much greater coefficient of thermal expansion. Again, these shrinkage stresses will relax with time.

3.6. Cyclic Loads

In water distribution systems, pressure variations are introduced by changes in the total demand for water (61). However, such variations only occur to a noticeable extent at peak demand periods so that, overall, the pressure remains relatively constant. In addition, the effects of fatigue in service for PE pipe systems are not considered to be a problem (62, 63).

4. PERFORMANCE ASSESSMENT OF BUTT FUSION WELDS IN POLYETHYLENE PIPES

4.1. Introduction

There are two types of tests which are used to assess the performance of butt fusion welds in PE pipes: those tests carried out on the whole product (large-scale tests) and those tests carried out on coupons cut from the pipe joint (coupon tests).

There are advantages and disadvantages to both types of tests. Large-scale tests are expensive but can be more representative of the conditions seen in service, whereas coupon tests are normally quick and inexpensive to perform but do not necessarily give relevant information regarding the performance of the weld in service.

4.2. Failure Modes

4.2.1. Fracture behaviour

A large number of empirical tests have been developed which can measure the impact strength of welds in polymers (64). As in metals, however, the properties measured by different test techniques are difficult to correlate.

In general, polymers exhibit a transition from tough ductile behaviour to low energy brittle failure as either strain rate is increased or temperature is decreased. Ductile fracture is generally associated with extensive shear yielding, a process which is thought to arise from the crystalline regions in the polymer undergoing processes such as slip or twinning (64).

The predominant initiating mechanism for brittle fracture is crazing. This is a cavitation process in which microvoids are nucleated under an applied tensile stress but do not coalesce to form a crack because they are bridged by fibrils spanning the craze (65). Crack propagation occurs by rupture of these fibrils (66, 67). Crazing has no metallic failure mode analogue; polymeric materials do not possess a failure mode analogous to cleavage. The chemical environment is a particularly important factor influencing both the failure mechanism and toughness.

Charpy testing has been shown to be suitable for comparing the toughness of welded joints in plastics (68). However, it does not allow a quantitative relationship to be developed between tolerable flaw size and applied stress. For this reason, linear elastic fracture mechanics (LEFM) has been investigated as a technique for characterising the toughness of plastics welds (see section 4.4.4).

4.2.2. Fatigue behaviour

The principal variable controlling fatigue behaviour is stress range, but for polymers a number of other factors are also important which are relatively insignificant in metals. Environment and ambient temperature strongly influence behaviour, but cyclic frequency also plays a large role. The reason for this is in the viscoelastic nature of polymers which exhibit mechanical hysteresis even at relatively low strains. Hence, under cyclic loading, energy is dissipated as heat, resulting in an increase in specimen temperature which can rise significantly above ambient. This can influence fatigue performance and hence, to obtain appropriate fatigue data, the test conditions, and in particular the cycling frequency, must closely parallel the service conditions.

Fatigue failure can be separated into an initiation phase and a propagation phase – in the presence of flaws the propagation phase dominates behaviour and it can be conservatively assumed that the initiation phase is absent. The mechanism of crack advance is material and stress dependent, but generally is believed to involve damage accumulation ahead of the crack tip by a combination of shear yielding and crazing.

As in metals, the rate of fatigue crack extension per cycle (da/dN) can be expressed in terms of the stress intensity factor range (ΔK):–

$$\frac{da}{dN} = A(\Delta K)^m \qquad [10]$$

where A and m are constants depending on material, temperature, frequency, stress ratio and environment. The parallel with the behaviour of metals extends to the low ΔK regime where a threshold stress intensity factor (ΔK_t) is observed, below which fatigue crack extension does not occur.

4.2.3. Stress rupture behaviour

When a welded PE specimen is subjected to a stress it will fail after a certain length of time, depending on the magnitude of the stress, and a log–log graph of stress versus time to failure can be produced. A schematic graph is shown in Fig. 13. As can be seen there are three distinctive regions.

At high stresses and short times the failure occurs by a local, large–scale deformation. This is referred to as 'ductile' failure and is due to shear yielding.

At lower stresses and longer times the failure occurs where no visible large–scale deformation is apparent. This is referred to as a 'brittle' failure and is due to a crazing process known as 'slow crack growth' (SCG), a phenomenon based on the initiation of a crack from a defect in the specimen which eventually propagates through the thickness.

At even lower stresses, and correspondingly longer times, the specimen will fail due to degradation of the PE (69). This type of failure is only important at high service temperatures (70) or in environments which will accelerate degradation (69). For normal distribution of water and gas, failure due to degradation is not a problem. This type of failure, therefore, will not be mentioned further in this report.

The lifetime of a PE pipe system in service, and whether it will fail in a ductile manner or via SCG, depends on the maximum local stress (see section 3) and the position of the 'knee' between ductile and brittle failure modes (Fig. 13).

4.3. Large–Scale Tests

The integrity of pressure pipeline systems depends not only on the extruded pipe but also on the components in the system, namely joints and fittings. To take account of possible differences between the performance of the pipe and the components, industry tends simply to increase the safety factor, CD. Clearly then, there is a need to assess the long–term performance of joints in order to define a safety factor which is not over–estimated.

In most international specifications the main test for ensuring a 50 year lifetime for PE pipe systems is the internal hydrostatic pressure test (4–6), where pipe samples, of length typically greater than three times the outside diameter, are fitted with end caps designed to make a pressure–tight connection between the pipe sample and the pressurising equipment. The pipe is then filled with water (via the end cap) and immersed unconstrained in a water bath, which is maintained at a constant temperature, and after a conditioning period (typically 1 hour) the hydrostatic pressure is applied and the time to failure measured. As stated in the previous section the lifetime of a PE pipe system depends on two properties: the long–term strength and the resistance to SCG. For today's PE pipe systems, SCG is regarded as the more important long–term failure mode. However, a test at a temperature of 20°C will only measure the pipes' long–term ductile strength. Therefore, a test procedure is required which accelerates the brittle SCG mode. The method most commonly used is to pressure test at elevated temperatures (8–10, 71). This results in a family of stress rupture curves as shown in Fig. 14, from which the shape of the 20°C curve can be extrapolated to 50 years using various mathematical techniques (7, 72–76).

The use of this technique has been criticised (77, 78) due to the fact that the crystallinity, and therefore properties, of PE change when subjected to elevated temperatures. However, the use of vertical and horizontal shift functions to produce a coherent master curve (76), has shown that these effects can be accounted for, at least for test temperatures of up to 80°C.

The hydrostatic pressure test also forms the basis of quality control and type tests which are normally performed at a temperature of 80°C and for periods of between 120 and 1000 hours, depending on the hydrostatic test pressure used (9, 10, 71, 79).

These tests, however, cannot be regarded as adequate methods for assessing the performance of butt fusion welds since joints are more vulnerable to stresses in the axial direction. In the hydrostatic pressure tests the axial stress in the pipe wall, due to the pressure against the end caps, is only half the hoop stress (see section 1.2.3) and therefore, except in the case of very poor quality welds, failure always occurs in the pipe wall first (17). Although these tests suggest that the weld is as strong as the pipe, the only real information they give is that the weld has a strength greater than 50% of that of the pipe. This is an insufficient criterion since, in service, a pressure pipe has to withstand additional axial stress due to bending, thermal contraction, misalignment, etc. (see section 3).

In order to give a better indication of the structural integrity of the weld in service, a number of standards also include the testing of a pipe with a radial offset at the weld of at least 10% of the wall thickness (10, 71).

The effect of misalignment in butt fusion welds on the performance of small diameter (≤125mm) PE pipe systems, during either constant or cyclic pressure tests at 80°C, has been studied extensively by Bowman et al (30, 31, 80, 81). They have shown that, for a radial offset greater than about 10% of the wall thickness, failure occurs in a shorter time than for pipes containing perfectly aligned butt fusion welds and always occurs at the weld, the plane of fracture being circumferential and initiating from the notch formed between the weld bead and the internal bore of the pipe.

Although this work showed that misalignment in butt fusion welds can have a significant effect on the life of the joint it was not possible to extrapolate the data back to predict the performance of perfectly aligned joints. This was because misalignment not only increases the axial stress of the weld but also sharpens the notch between the weld bead and pipe wall.

Butt fusion welded PE pipes have also been assessed using internal pressure fatigue tests (30, 80). Results showed that, for 125mm OD pipes, failure occurred away from the weld, whereas for smaller sizes a number of failures occurred at the joint, due to either circumferential or axial cracking. Although these test generated data on the failure of butt fusion welds rather than pipes, at least for small diameters, this information was only relevant for assessing the performance of welds under fatigue loading.

Therefore, at present there is still no large–scale test available to predict the service life under constant pressure of well–made butt fusion welds in PE pipes.

4.4. Coupon Tests

4.4.1. Tensile tests

Tensile testing is probably the most common method for determining the quality of PE welds (18, 19, 23, 28, 78, 82–89).

A generally accepted approach is to use a dumb–bell specimen, such as the one shown in Fig. 15, to determine a 'welding factor', *f*, (90) given by:-

$$f = \frac{\textit{yield stress of weld}}{\textit{yield stress of parent material}} \qquad [11]$$

where the yield stress is defined according to BS 2782: Part3: Method 321 (91). According to Deutscher Verband Für Schweisstechnik (DVS), acceptable values of *f* for PE butt fusion welds are 0.9 and greater (92).

There are two main problems associated with using test specimens such as the one shown in Fig. 15: 1) it is not ideal for curved samples, such as those taken from pipes and 2) except for very poor welds, the welding factor, *f*, is very close to unity and fracture tends to occur remote from the weld (17).

For these reasons, waisted sample geometries (such as the one shown in Fig. 16), which force fracture to occur at the weld, are often used for pipe welds (10, 87, 89). However, even with this geometry, values of *f* are relatively insensitive to changes in welding conditions (19, 82, 93). For this reason the UK Water Industry does not specify a required welding factor. Instead, it specifies only that the fracture surfaces should be ductile (10, 94). The problem with this approach is that it is a qualitative assessment of the weld quality and is based on the experience of the examiner.

It has been suggested that a better indication of weld quality may be elongation to break (18, 36, 82) since welds with a welding factor of 1.0 only achieve a maximum elongation to break of around 50% of that of the parent pipe.

It has also been suggested that the energy to break per cross-sectional area can be used to quantify the weld quality (93), acceptable values being greater than 50% of that of the parent pipe tested with the same specimen geometry.

Other workers have suggested that the long-term welding factor may be predicted from a short-term tensile test performed at a sufficiently low temperature to cause brittle failure (95). Such tests have been performed (86) using another waisted test specimen geometry (see Fig. 17) but results have not been correlated with long-term data.

4.4.2. Bend tests

Probably the simplest test for evaluating the quality of butt fusion welds in PE is a bend test (96–98). Rectangular section unnotched test pieces, cut from across the butt fusion weld, are loaded in three point bend and the angle at which either fracture occurs or a crack initiates is measured. If no cracking of the weld occurs the test is terminated at a bend angle of 160°.

It has been shown (99) that more useful information can be obtained from this test by performing it at low temperatures. However, results could not be correlated with the long-term properties of the weld.

4.4.3. Impact tests

Impact testing has been another approach used to estimate the fracture toughness of butt fusion welds in PE pipes (18, 68, 93, 100–102).

There are basically two types of impact test, namely flexural tests, such as Charpy (103–105), Izod (105, 106) and falling weight (101), where a striker impacts a notched bar (see Fig. 18); and the tensile impact test (100), where a dumb-bell specimen is fractured in tension (see Fig. 19).

Conventional flexural impact testing of plastics measures the total energy absorbed in fracturing a standard blunt notched (0.25mm root radius) specimen (18, 104) (see Fig. 20). However, the values obtained are heavily dependent on specimen geometry and test conditions, and the properties measured by different test techniques are difficult to correlate. Therefore, such tests provide only a crude means of characterising fracture toughness.

An alternative approach is to use LEFM (107) by incorporating a much sharper crack (10–20μm) in the specimen. This technique involves the introduction of a sharp crack by a razor blade into the standard specimen, after which the energy required to fracture the specimen (U_c) is measured in the normal manner. By varying the depth of the initial crack, a number of values of fracture energy may be obtained.

According to LEFM principles the critical strain energy release rate, G_c (a measure of fracture toughness) is given by:–

$$G_c = \frac{U_c}{BD\Phi} \qquad [12]$$

where ϕ is a geometric factor (107), B is the specimen thickness and D is the width.

Hence, by plotting U_c against $BD\phi$ a straight line of slope G_c should be obtained.

One of the major problems when flexural impact testing PE welds is the positioning of the notch. Ideally the notch tip should be at the weld interface. If the weld bead is symmetrical, it is possible to use the centre of the bead as an accurate locator of the interface. However, this is not the case if the bead is unsymmetrical. This has been overcome by introducing a notch into the joint (using PTFE tape) during the welding process (102).

Studies of the impact performance of butt fusion welds in PE pipes using both flexural and tensile impact tests have revealed that these tests are more useful than tensile tests for examining the effect of changing welding conditions on the quality of the joint (108, 109). Indeed, joints with a measured welding factor of 1.0 from a tensile test can have impact values less than 50% of the parent pipe (18, 93, 102, 109). It has also been found that more consistent results are obtained when the weld beads are removed (68).

4.4.4. Long–term tests

Long–term coupon tests on specimens taken from butt fusion welds in PE pipes are used primarily to compare the resistance to SCG of the joint with that of the parent pipe (110–112). The constant tensile load test consists of subjecting dumb–bell specimens, cut from the butt fusion weld and from the parent pipe, to a range of constant tensile loads, under defined conditions of temperature and environment (see Fig. 21). The times to failure are then measured and creep rupture curves of log (stress) against log (time to failure) are generated. From these curves (Fig. 22) a long–term tensile welding factor can be determined (ie. the ratio of the stress levels which result in equal times to failure for the parent and welded specimen). According to DVS (92), a butt fusion weld in PE pipe should have a long–term tensile welding factor of at least 0.8 to be acceptable. However, the calculated values of long–term tensile welding factor have been shown to be dependent on the geometry of the bead and specimen (27).

An important aspect of the long–term testing of small–scale specimens is the possibility of predicting the lifetime of a pipe system without having to perform large–scale, expensive tests on the pipes themselves. This has been attempted using a fracture mechanics methodology for assessing the failure of butt fusion joints in HDPE pipe due to SCG (113–115). Three point bend specimens containing a centrally located notched joint were tested at ambient and elevated temperature for several combinations of load and initial crack size. A viscoelastic model was then used to deduce the crack growth history from the measured load-point displacement. This work has led to the production of master curves for the correlation of both initiation time and crack growth rate with stress intensity factor. By using these master curves, the effect of a crack-like defect on the total lifetime could be estimated.

However, the above methodology is only valid for materials which exhibit sufficient linear elastic fracture behaviour. Unfortunately, this is not the case for current PE pipe materials. Nevertheless, this work has indicated that the resistance to slow crack growth of butt fusion joints can be significantly less than that of the parent material, even though the joint and parent material may have similar short-term tensile strengths (114), a finding which has also been suggested from elevated temperature tensile creep tests (2).

5. DISCUSSION

At present there are conflicting views regarding the optimum butt fusion welding procedure for large diameter PE pipes and it is difficult to ensure similar weld quality to that of small diameter pipes. This, together with the fact that the likelihood of brittle fracture from a defect increases with pipe wall thickness (36), means that it is essential that reliable procedures are developed to assess the structural integrity of welded joints in PE pipe systems. Such procedures would determine the relationship between the size and position of any defects, the service conditions and the properties of the material and, combined with reliable NDT techniques, would mean that on-site pipe system lifetime predictions could be made. This, in turn, would lead to greater confidence in the use of butt fusion welded PE pipes in more safety critical applications and to the adoption of more realistic safety factors.

To predict the service life of butt fusion welds in PE pipes, requires that long-term tests are performed under conditions which allow reliable extrapolation of results back to those conditions which occur in service. Long-term tests on coupons cut from the weld may give some indication of the service life of a joint, but the stresses in a coupon will be different to those in the whole welded pipe and results from coupon tests will need to be correlated with results from large-scale tests.

Large-scale hydrostatic pressure testing of whole butt-fusion welded pipe sections may be suitable for predicting the service life of the parent pipe. However, it is inadequate for predicting the service life of pipe butt fusion welds since, invariably, failure occurs in the pipe wall before the welded joint, giving no real information about the long-term quality of the weld. Hydrostatic pressure tests on welded pipe samples with a radial offset at the joint generate failure at the weld. However, the misalignment produces an artificially sharp notch between the weld bead and the pipe wall which may not allow reliable extrapolation back to perfectly aligned joints.

Therefore, at present there is no test available which has been shown conclusively to predict the service life of well-made butt fusion welds in PE pipes, or conversely, to estimate the maximum allowable flaw size for a specified lifetime and loading. Consequently, it is not known whether present NDT techniques are sufficiently sensitive to achieve on-site pipeline evaluation.

There are a number of short-term coupon tests commonly used for QA purposes and also to optimise the butt fusion welding process. However, the criteria that a 'good quality' weld should achieve is different for different tests. For example, one tensile test (10) defines a good quality weld as one which fails in a ductile manner; another tensile test requires a welding factor of between 0.9 and 1.0 (92); whereas in impact tests identical welds can only achieve a 'welding factor' of around 0.5 (18, 93, 102, 109).

In reality, for butt fusion welds in PE pipes the most important criterion is that the weld should have an adequate service life. However, since PE is viscoelastic, and hence prone to time-dependent failure, these short-term tests are unable to quantify the long-term performance of the joint, and insufficient work has been carried out to correlate the results from short-term tests and long-term tests.

6. CONCLUSIONS

This report has reviewed the present position of the PE pipes industry in terms of assessment of the structural integrity of butt fusion welded joints. The following conclusions can be drawn:

- As the use of PE for gas and water pipe systems increases, the need for reliable procedures to determine the structural integrity of welded PE pipe systems is becoming more and more important. This is particularly the case for larger diameter pipes, where optimum welding conditions are still being developed.

- In order to determine the structural integrity of welds in PE pipe systems in service three factors must be known: the existence of any defects, the service conditions and the properties of the weld. In addition, the relationship between these three factors must also be known.

- Although NDT techniques can detect a number of flaws in welded PE pipe systems, including surface damage and volumetric flaws, the relationship between flaw size and pipe lifetime has not been determined. Consequently it is not known whether the flaws which have been detected are of a critical size or not.

- There are a number of possible stresses in welded PE pipe systems, including those caused by internal pressure, residual stress, misalignment, bending and thermal contraction, and all of these must be accounted for if realistic lifetime predictions are to be made.

- The current tests available for welded PE pipe systems can be divided into two: those that test specimens cut from the joint and those that test whole pipe sections. Tests on whole pipe sections more closely simulate service conditions but current hydrostatic pressure tests, even with radial misalignment at the joint, do not account for axial stresses due to bending and thermal contraction and cannot therefore be used to predict weld service life reliably. Tests on specimens cut from the joint are normally cheaper and easier to perform, but bear little resemblance to service conditions.

- The use of fracture mechanics may be capable of generating lifetime data for some PE pipes and pipe welds but its validity for the very tough PE grades currently used in Europe has not yet been established.

- The 'quality' of a weld as measured by a particular test (whether large-scale or coupon, short-term or long-term) depends upon the test itself. A number of tests (eg. standard hydrostatic pressure test and tensile tests) indicate that a well-made weld is as strong as the pipe. Other tests, such as impact tests, indicate that the same weld is only around half as strong as the pipe. This obviously begs the question "which test gives the correct results?".

- There is a need for a new large-scale test which can reliably predict the service life of butt fusion welds in PE pipes.

- There is also a need to compare results from the many coupon tests currently used in industry and to correlate them with the results from a test which can generate reliable weld service life predictions. This will highlight which of the coupon tests gives the best indication of the long-term durability of the weld and should therefore be used for QA purposes and to optimise butt fusion welding parameters.

7. ACKNOWLEDGEMENTS

The authors would like to express their thanks to their colleagues at TWI for their generous encouragement and helpful suggestions during the preparation of this report. The work was carried out within the core research programme of TWI which is jointly funded by the Information and Manufacturing Technology Division of the Department of Trade and Industry, and the Industrial Members of TWI.

8. REFERENCES

1. BS PD 6493: 1991, 'Guidance on methods for assessing the acceptability of flaws in fusion welded structures'.

2. JANSON L-E, 'Plastic pipes for water supply and sewage disposal', Neste, Stockholm, 1989.

3. ANTONIETTI G, COSTA G and PERI F, 'Developments in the welding of PE 50 polyethylene pipes and in the relevant qualifications', Weld. Int., 5, (1988), 432.

4. BS 4728:1971, 'Method for determination of the resistance to constant internal pressure of thermoplastics pipe'.

5. ISO 1167:1973, 'Plastics pipes for the transport of fluids – determination of the resistance to internal pressure'.

6. ASTM D1598–81, 'Time–to–failure of plastic pipe under constant internal pressure'.

7. ISO/DTR 9080.2 (1989), 'Plastics pipes for the transport of fluids – standard extrapolation method for the long term resistance to constant internal pressure'.

8. CEN/TC 155: System Standard 20: Final Draft, 'Plastics piping systems and ducting systems – polyethylene (PE) piping systems for water supply'.

9. BS 7281:1990, 'Polyethylene pipes for the supply of gaseous fuels'.

10. WIS No. 4–32–03. May 1987. (Amendment July 1989), 'Specification for blue polyethylene (PE) pressure pipe for cold potable water (nominal sizes 90 to 1000 for underground or protected use)'.

11. GREIG J M and EWING L, 'Development and application of polyethylene gas pipe systems for use at pressures beyond 7 bar', 8th Int. Conf on Plastics Pipes, Koningshof, The Netherlands, 21–24 September 1992.

12. IGE/TD/4 (1981), Edition 2, 'Gas services'.

13. IGE/TD/3 (1983), Edition 2, 'Distribution mains'.

14. DODIN M G, 'Welding mechanisms of plastics: a review', J Adhesion. 12 (1981), 99.

15. POTENTE H and REINKE M, ' Welding parameters and properties of polyolefin parts', Plast. and Rubb. Proc. and Appl., 1, (1981), 149.

16. WIS No 4–32–08, August 1988, ' Specification for site fusion jointing of MDPE pipe and fittings'.

17. WOLTERS M and VENEMA B, 'Mechanical quality of welds in polyethylene pipe systems', 5th Int. Conf. on Plastics Pipes, University of York, 8–10 September 1982.

18. WATSON M N, 'A study of the butt fusion (hot plate) welding process for polyethylene pipe', TWI Report 382/1988.

19. WATSON M N, 'The butt fusion jointing process – fundamental aspects', 7th Int. Conf. on Plastics Pipes, University of Bath, 20–22 September 1988.

20. PIMPUTKAR S M, 'The dependence of butt fusion bond strength on joining conditions for polyethylene pipe', ANTEC '89, New York, 1–4 May 1989.

21. WATSON M N and MURCH M G, 'Recent developments in hot plate welding of thermoplastics, ' Poly. Eng. Sci, 29 (19), (1989), 1382.

22. WATSON M N, 'The butt fusion (hot plate) welding process for PE and PP pipe – improvements to productivity', TWI Report 419/1990.

23. MARSHALL G P, 'The influence of welding parameters on the toughness of butt fusion welds in MDPE', Advances in Joining Plastics and Composites, Bradford, 10–12 June 1991.

24. BRIDGSTOCK E, WATSON M N and BOWMAN J, 'Attaining ISO 9000 series quality levels on-site via computer controlled butt fusion machines', 13th Plastic Fuel Gas Pipe Symp., San Antonio, USA, 1–4 November 1993.

25. CARTWRIGHT R, TROUGHTON M J, HINCHCLIFF F A, BOWMAN J and JENKINS P, 'High productivity butt fusion joining of polyethylene pipes', 9th Int. Conf. on Plastics Pipes, Edinburgh, 18–21 September 1995.

26. MORRIS J D and GILL B W, 'Fusion jointing of PE pipes: A water industry perspective', 8th Int. Conf. on Plastics Pipes, Koningshof, The Netherlands, 21–24 September 1992.

27. VANSPEYBROECK P, 'Analysis of results of the constant tensile load test for fused PE gas pipe materials', 9th Int. Conf. on Plastics Pipes, Edinburgh, 18–21 September 1995.

28. BARBER P and ATKINSON J R, 'The use of tensile tests to determine the optimum conditions for butt fusion welding certain grades of polyethylene, polybutene-1 and polypropylene pipes', J Mater. Sci, 9 (1974), 1456.

29. BARTON S J and CHERRY B W, ' Butt fusion welding of high density polyethylene pipe', Adhesion 7: 20th Ann. Conf. on Adhesion and Adhesives, London (1983).

30. PARMAR R and BOWMAN J, 'Crack initiation and propagation paths for brittle failures in aligned and misaligned pipe butt fusion joints', Poly. Eng. Sci, 29 (19), (1989), 1396.

31. PARMAR R and BOWMAN J, 'The failure of butt fusion joints in polyethylene pipe systems', 6th Int. Conf. on Plastics Pipes, University of York, 26–28 March 1985.

32. MAINE L and STAFFORD T W, 'PE jointing techniques', 6th Int. Conf. on Plastics Pipes, University of York, 26–28 March 1985.

33. EDWARDS G R, 'The development of non–destructive test procedures for hot plate butt fusion welds in polyethylene', TWI Report 370/1988.

34. SIMS J, 'The butt fusion (hot plate) welding process for polyethylene pipe for 125 and 400mm diameter'. TWI Report 434/1991.

35. GIRARDI M A, 'The evaluation of mechanical testing techniques and the influence of defects in MDPE butt fusion welds', 8th Int. Conf. on Plastics Pipes, Koningshof, The Netherlands, 21–24 September 1992.

36. MURCH M G and TROUGHTON M J, 'A study of the applicability of the tensile weld test for thick walled polyethylene pipe', TWI Report 533/1995.

37. LUMB R F, 'Four ways to test PE joints', Pipeline Industry, 4 (37), (1972), 47.

38. MAINE L and STAFFORD T G, 'In field quality control of fusion jointing methods', 7th Int. Conf. on Plastics Pipes, University of Bath, 19–22 September 1988.

39. prEN 1555–6, 'Plastics piping systems for gaseous fuels supply – polyethylene (PE) – Part 6: recommended practice for installation', 1995.

40. McCLINTON M, 'How to detect flaws in PE pipe butt fusion joints', Pipe Line Industry, November 1986.

41. HOUSE L, MANK J, PETTLINSKI T and WILLIAMS W, 'Quality assurance of heat fused polyethylene plastic pipe joints – phase III', GRI Report 85/0284, January 1986.

42. AHMED H, 'Use of ultrasonic tests to check PE pipe joints', Pipe Line Industry, 70(1), (1989), 19.

43. STRIPLIN T H, 'Ultrasonic evaluation of polyethylene butt fused joints', Buried Plastic Pipe Technology, ASTM STP 1093, 1990.

44. LUMB R F, BENNETT A and WARD C R, 'A review of NDT at British Gas Engineering Research Station', BJNDT, September 1980, pp217–225.

45. WERNER ROYE, ' X–ray testing of thermoplastics welding joints', 4th European Conf. on NDT, London 1987.

46. ANON, 'X–rays show up pipe fitting flaws', Brit. Plast. Rubb., Feb 1985, 47.

47. POTENTE H and ZEEUW K, 'Non-destructive testing of heated tool butt welded pipes using optical methods', 4th Int. Conf. on Plastics Pipes, University of Sussex, 28–30 March 1979.

48. DEAN G D, ENNOS A E and QUINN G B, 'Detection of cracks in plastic pipe joint by reflection holography'. Plast. and Rubb. Proc. and Appl., 3(5), (1983), 229.

49. ESAULENKO G B, KONDRATENKO V YU and BEZRUK L I, 'Study of strength characteristics of polyethylene butt welded joints and development of testing methods', Int. Inst of Welding, Document No XVI–259–90, (1990).

50. ANON, 'Total pipe systems move to polyethylene for potable water, gas, sewerage and industrial applications', Stewarts and Lloyds Plastics (1988).

51. ESAULENKO G B, KONDRATENKO V YU, BEZRUK L I and KORAB G N, 'On the possibility of using the morphological and mechanical properties of fin from polyolefin butt welded joints to estimate their quality', Int. Inst. of Welding, Document No XVI–1011–88, (1988).

52. GEORGIOU G A and MACDONALD I A, 'Ultrasonic and radiographic NDT of butt fusion joints in polyethylene', TWI Report 465/1993.

53. HOUSE L J and LUSTIGER A, 'Automated ultrasonic inspection of polyethylene butt fusion joints', 8th Plastic Fuel Gas Pipe Symposium, New Orleans, USA, 1983.

54. VENEMA B and WOLTERS M, 'Evaluation of possibilities for non-destructive inspection of butt fusion joints in PE piping systems', Advances in Joining Plastics and Composites, Bradford, 10–12 June 1991.

55. MUNNS I J and GEORGIOU G A, 'Ultrasonic and radiographic NDT of butt fusion welds in polyethylene pipe', 9th Int. Conf. on Plastics Pipes, Edinburgh, 18–21 September 1995.

56. INGEN HOUSZ J F, 'Residual stress in internally coded pipe', 5th Int. Conf. on Plastics Pipes, University of York, 8–10 September 1982.

57. KUHLMAN C J, TWEEDY L K and KANNINEN M F, 'Forecasting the long-term service performance of polyethylene gas distribution pipes', 8th Int. Conf. on Plastics Pipes, Koningshof, The Netherlands, 21–24 September 1992.

58. BEECH S H, BURLEY C and BUNN H C, 'Residual stress in large diameter MDPE water pipe', 7th Int. Conf. on Plastics Pipes, University of Bath, 19–22 September 1988.

59. REINKE M and POTENTE H, 'The effects of the processing history on the internal properties of HDPE pipes and their butt-welding and how they are affected by annealing', ANTEC '82, 59.

60. MADDOX S J, 'Fitness–for–purpose assessment of misalignment in transverse butt welds subject to fatigue loading', TWI Report 279/1985.

61. KIRBY P C, 'Surge and fatigue in unplasticized PVC sewer rising mains', Plast. and Rubb. Mater. and Appl., May 1980, 78.

62. DUKES B W, 'The dynamic fatigue behaviour of uPVC pressure pipe', 6th Int. Conf. on Plastics Pipes, University of York, 26–28 March 1985.

63. BEECH S H, HEADFORD A, HUNT S and SANDILANDS G, 'The resistance of polyethylene water pipeline systems to surge pressure', 9th Int. Conf. on Plastics Pipes, Edinburgh, 18–21 September 1995.

64. KINLOCH A J and YOUNG R J, 'Fracture Behaviour of Polymers', Elsevier, London, (1983).

65. LEIS B N, 'Analysis of enhanced resistance to slow crack growth in PE materials: implications for performance prediction,' 8th Int. Conf. on Plastics Pipes, Koningshof, The Netherlands, 21–24 September 1992.

66. LU X, WANG X and BROWN N, 'Slow fracture in a homopolymer and copolymer of polyethylene', J. Mater. Sci., 23, (1988), 643.

67. BARRY D and DELATYCKI O, 'Static fatigue fracture of polyethylene: a morphological analysis', J. Poly. Sci.: Part B: Poly. Phys., 25, (1987), 883.

68. HALE G E, 'Assessment of mechanical property and fracture mechanics tests applicable to plain material and welded joints in thermoplastics', TWI Report 383/1988.

69. IFWARSON M and LEIJSTROM H, 'What controls the lifetime of plastic pipes and how can the lifetime be extrapolated', 8th Int. Conf. on Plastics Pipes, Koningshof, The Netherlands, 21–24 September 1992.

70. GEDDE U W, SMITH G D, IFWARSON M and VIEBKE J, 'Antioxidant loss in hot–water polyolefin pipes: experimental data and modelling', 8th Int. Conf. on Plastics Pipes, Koningshof, The Netherlands, 21–24 September 1992.

71. BGC/PS/PL2: Part 1, 'Specification for polyethylene (PE) pipes and fittings for natural gas and suitable manufactured gas – pipes', June 1986.

72. LARSON F R and MILLER J, 'A time temperature relationship for rupture–creep stresses', Trans. Am. Soc. Mech. Eng., (1952), 765.

73. BRAGAW C G, 'The forecast of polyethylene pipe and fittings burst life using rate process theory', 5th Int. Conf. on Plastics Pipes, University of York, 8–10 September 1982.

74. DE PUTTER W J, 'Extrapolation of stress rupture data', 7th Int. Conf. on Plastics Pipes, University of Bath, 19–22 September 1988.

75. ASTM D2837–90, 'Standard method for obtaining hydrostatic design basis for thermoplastic pipe materials'.

76. POPELAR C F, POPELAR C H and KENNER V H, 'Viscoelastic material characterization and modelling for polyethylene', Poly. Eng. Sci., 30(10), (1990), 577.

77. SCHONEFELD G and WINTERGERST S, 'Effects of heat–treatment upon the density, internal stress and bursting strength, of drinking water pipes constructed from plasticized and rigid polyethylene', Kunststoffe, 61, (1971), 33.

78. ATKINSON J R and DECOURCY D R, 'The assessment of fusion joint quality and some evidence for the presence of molecular orientation in butt fusion welds in polyethylene pipes', Plast. and Rubb. Proc. and Appl., 1 (4), (1981), 287.

79. JIS K6774: 1989, 'Polyethylene pipes for the supply of gaseous fuels'.

80. BOWMAN J, 'The fatigue behaviour of medium density polyethylene pipe systems', 7th Int. Conf. on Plastics Pipes, University of Bath, 19–22 September 1988.

81. BOWMAN J and PARMAR R, 'The importance of axial misalignment on the long–term strength of polyethylene pipe butt fusion joints', Poly. Eng. Sci., 29(19), (1989), 1406.

82. DECOURCY D R and ATKINSON J R, 'The use of tensile tests to determine the optimum conditions for butt welding polyethylene pipes of different melt flow index', J. Mater. Sci., 12, (1977), 1535.

83. POTENTE H and TAPPE P 'Heated tool–butt welding of polyethylene pipes – parameters and testing techniques,' Mat. and Des., 5, (1985), 273.

84. DVS 2203: Part 2: 1985, 'Testing of welded joints of thermoplastics – tensile test'.

85. CNS A 89–801, 'Tentative standard: welding and allied processes: transverse tensile test on thermoplastic butt welds', August 1990.

86. TAYLOR N, KLAIBER F and WERMELINGER J, 'IR welding afforded promising future', TWI Bulletin 4, 34, July/August 1993.

87. CEN/TC249/SC5/WG2 N14, 'Testing of welded joints of thermoplastics – low temperature tensile test', Feb 1994.

88. CEN/TC249/SC5/WG2 N49, 'Testing of welded joints of thermoplastics semi–finished products – Part 2: tensile test', March 1996.

89. ISO CD 13953, 'Tensile testing on specimens from butt fused polyethylene (PE) pipe assemblies'.

90. BARBER P and ATKINSON J R, 'Some microstructural features of the welds in butt-welded polyethylene and polybutene-1 pipes', J. Mater. Sci., 7, (1972), 1131.

91. BS 2782: Part 3: Method 321:1994, 'Methods of testing plastics. Part 3 - mechanical properties, Method 321 - determination of tensile properties. General principles'.

92. DVS 2203: Part 1: 1986, 'Testing of welded joints of thermoplastics - test methods - Requirements'.

93. WILSON K A, 'Verification of butt fusion weld quality in large diameter PE100 water pipes', 9th Int. Conf. on Plastics Pipes, Edinburgh, 18-21 September 1995.

94. WIS No. 4-32-13, March 1993, 'Specification for blue higher performance polyethylene, HPPE/PE100, pressure pipes, nominal size 90 to 1000, for underground or protected use for the conveyance of water intended for human consumption'.

95. MICHAEL P, 'Parameteroptimerung beim heizelementstumpschweissen am beispel von HDPE', Schweisstechnische Forchungsberichte, Band 18, DVS, Dusseldorf, (1988).

96. DVS 2203: Part 5: 1985, 'Testing of welded joints of thermoplastics - technological bending test'.

97. CNS A 89-802, 'Welding and allied processes - transverse bend test on thermoplastic butt welds - flat and tubular assemblies, July 1991.

98. CEN/TC249/SC5/WG2 N38, 'Testing of welded joints of thermoplastics semi-finished products - Part 1: bend test', October 1995.

99. HESSEL J and JOHN P, 'Long-term strength of HDPE welded joints. Low-temperature flexural test used as evaluation criteria for butt welded heating elements', Kunststoffe 74 (1984), 7, 385.

100. DVS 2203: Part 3: 1985, 'Testing of welded joints of thermoplastics - tensile impact test'.

101. MCGRATH G C, 'The analysis of welded joints in plastic components using instrumented falling weight impact analysis and long term loading techniques', Advances in Joining Plastics and Composites, Bradford, 10-12 June 1991.

102. FOLKES M J, HAUNTON J and BOWMAN J, 'Microstructure development and its effect on the mechanical properties of butt fusion joints', Advances in Joining Plastics and Composites, Bradford, 10–12 June 1991.

103. DIN 53453: 1975, 'Testing of plastics: impact flexural test (May)'.

104. BS 2782: Part 3: Method 359:1984, 'Determination of Charpy impact strength of rigid materials (Charpy impact flexural test)'.

105. ASTM D 256: 1992, 'Standard test methods for impact resistance of plastics and electrical insulating materials'.

106. BS 2782: Part 3: Method 350:1984, 'Determination of Izod impact strength of rigid materials'.

107. PLATI E and WILLIAMS J G, 'The determination of the fracture parameters for polymers in impact', Poly. Eng. Sci, 15 (6), (1975), 470.

108. PIMPUTKAR S M, 'The dependence of butt fusion bond strength on joining conditions for polyethylene pipe', Poly. Eng. Sci., 29 (19), (1989), 1387.

109. NISHIMURA H and NARISAWA I, 'Evaluation of impact properties of butt–fusion–jointed medium density polyethylene pipes for gas distribution', Polymer, 32 (12), (1991), 2199.

110. DVS 2203: Part 4: 1986, 'Testing of welded joints of thermoplastics – tensile creep test'.

111. CEN/TC249/SC5/WG2 N50, 'Testing of welded joints of thermoplastics semi–finished products – Part 3: tensile creep test', March 1996.

112. ISO CD 13952, 'Static tensile test in tensio–active medium of butt fused polyethylene (PE) pipe joints'.

113. POPELAR C H, KENNER V H, POPELAR S F and PFEIL M C, 'Life expectancies for polyethylene butt fusion joints', 12th Plastic Fuel Gas Pipe Symp., Boston, USA, September 1991.

114. POPELAR S F, POPELAR C H and KENNER V H, 'Creep crack growth in joints of a thermoplastic', Int. J. Fracture, 50 (1991), 115.

115. PFEIL M C, KENNER V H and POPELAR C H, 'A fracture mechanics evaluation for the life expectancy of polyethylene butt fusion joints', Eng. Fracture Mech., 44 (1), (1993), 91.

TABLE 1 Main Applications for PE Materials

PE Type	Applications	
LDPE/LLDPE	Blown films	– bags, sacks, shrink wrap
	Blow moulded containers	– squeeze bottles
	Injection moulded articles	– toys, lids, household products
	Wire and cable insulation	
	Extrusion coating	
	Pipes	– small diameter
MDPE	Gas and water pipes	
	Film	
HDPE	Packaging film and carrier bags	
	Blow moulded bottles	– milk, household chemicals, toiletries
	Injection moulded articles	– margarine tubs, buckets, crates, toys
	Filament for ropes, nets, fabrics	

TABLE 2 History of PE Pipes (2)

Type of PE	First Produced	Pipe Diameters/mm
LDPE	1945	16–160
HDPE	1955	25–1600
MDPE	1971	25–1600
LLDPE	1986	16–160

TABLE 3 Classification of pipe materials

Minimum Required Strength/MPa	Class†	Designation††
6.3	A	PE63
8.0	B	PE80
10.0	C	PE100

† for gas pipes (9)
†† for water pipes (8)

TABLE 4 Pressure classification of pipes (8, 9)

Application	MRS/MPa	CD	HDS/MPa	Max Allowable Working Pressure, Pw/bar		
				SDR 11	SDR 17.6	SDR 26
Water	10.0	1.25	8.0	16.0	10.0	
	6.3	1.25	5.0	10.0	6.0	4.0
Gas	10.0	3.0	3.3	7.0	4.0	2.5
	8.0	3.0	2.7	5.5	3.0	2.0
	6.3	3.0	2.1	4.0	2.5	1.5

TABLE 5 Standard PE pipe dimensions (9)

Nominal outside diameter (d_e) (size)	Mean outside diameter		Wall thickness					
			SDR 11		SDR 17.6		SDR 26	
	min.	max.	min.	max.	min.	max.	min.	max.
mm	mm	mm	mm	mm	mm	mm	mm	mm
16	16	16.3	2.3	2.8	–	–	–	–
20	20	20.3	2.3	2.8	–	–	–	–
25	25	25.3	2.3	2.8	–	–	–	–
32	32	32.3	3.0	3.4	–	–	–	–
50	50	50.4	4.6	5.2	2.9	3.3	–	–
55	55	55.4	5.1	5.7	–	–	–	–
63	63	63.4	5.8	6.5	3.6	4.1	–	–
75	75	75.5	6.9	7.8	4.3	4.9	–	–
90	90	90.6	8.2	9.2	5.1	5.8	–	–
125	125	125.6	11.4	12.7	7.1	8.0	–	–
140	140	141.0	12.8	14.2	8.0	8.9	–	–
180	180	181.2	16.4	18.2	10.2	11.4	–	–
200	200	201.2	18.2	20.2	11.4	12.7	7.7	8.6
250	250	251.5	22.8	25.2	14.2	15.8	9.7	10.8
280	280	281.6	25.5	28.2	15.9	17.6	10.8	12.0
315	315	316.8	28.7	31.7	17.9	19.8	12.2	13.6
355	355	357.1	32.3	35.7	20.1	22.3	13.7	15.2
400	400	402.3	36.4	40.2	22.7	25.1	15.4	17.1
500	500	502.9	45.5	49.7	28.3	31.3	19.3	21.4

TABLE 6 Standard butt fusion weld bead widths (50)

Pipe Diameter/mm	Bead Width/mm
90–180	7–11
250	11–16
315–500	16–25

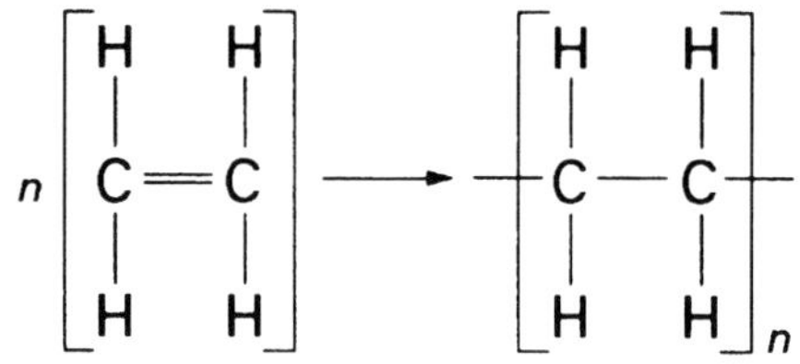

Fig. 1. Synthesis of polyethylene.

TYPE I Low Density	TYPE II Medium Density	TYPE III High Density	TYPE IV High Density
	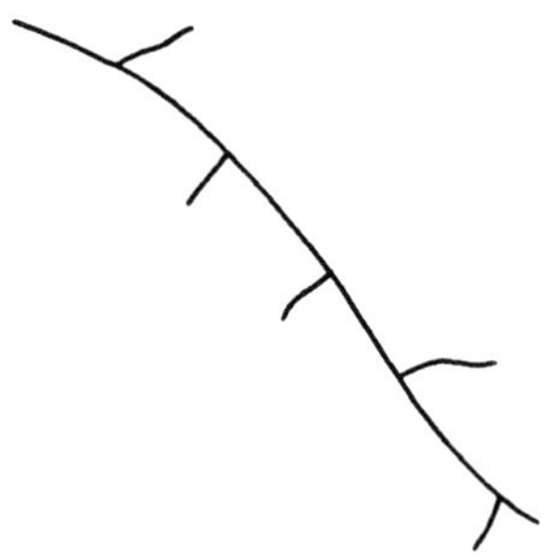	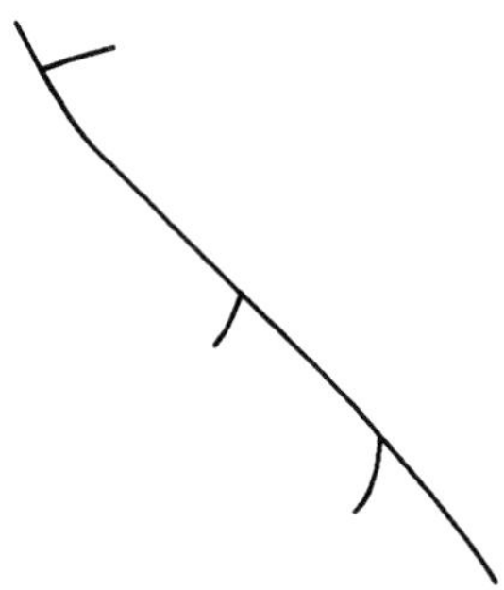	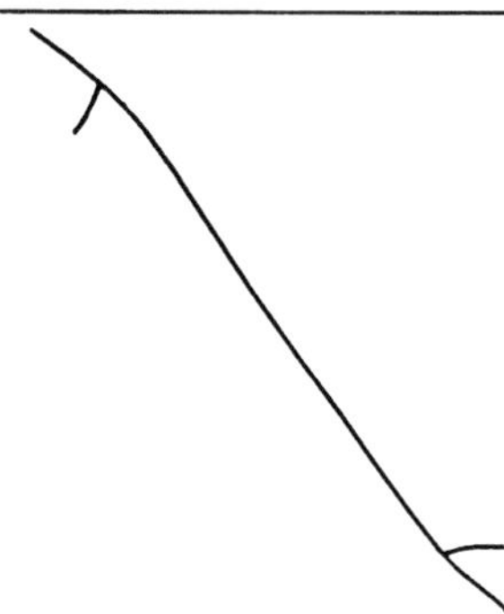
Polymer of ethylene	Copolymer of ethylene	Copolymer of ethylene	Polymer of ethylene
High pressure process	Low pressure process	Low pressure process	Low pressure process
Both long chain and short chain branches	Short chain branches only	Short chain branches only	Short chain branches only
Density: 0.910–0.925gcm^{-3}	Density: 0.926–0.940gcm^{-3}	Density: 0.941–0.959gcm^{-3}	Density: 0.960 + gcm^{-3}
>25 branches per 1000 carbon atoms	10–25 branches per 1000 carbon atoms	0.1–9 branches per 1000 carbon atoms	<0.1 branches per 1000 carbon atoms

Fig. 2. Types of polyethylene.

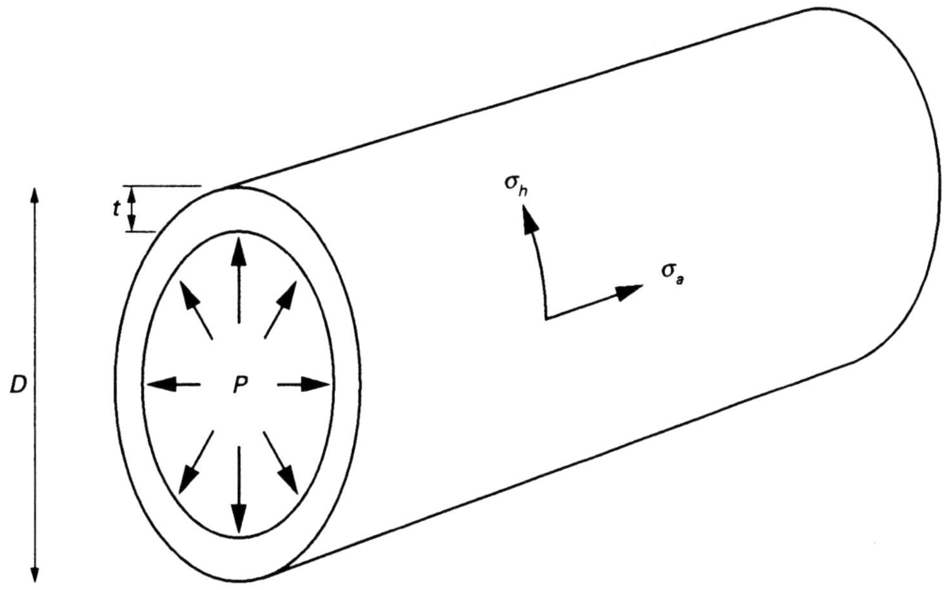

Fig. 3. Schematic diagram of stresses generated in wall of pipe under internal pressure.

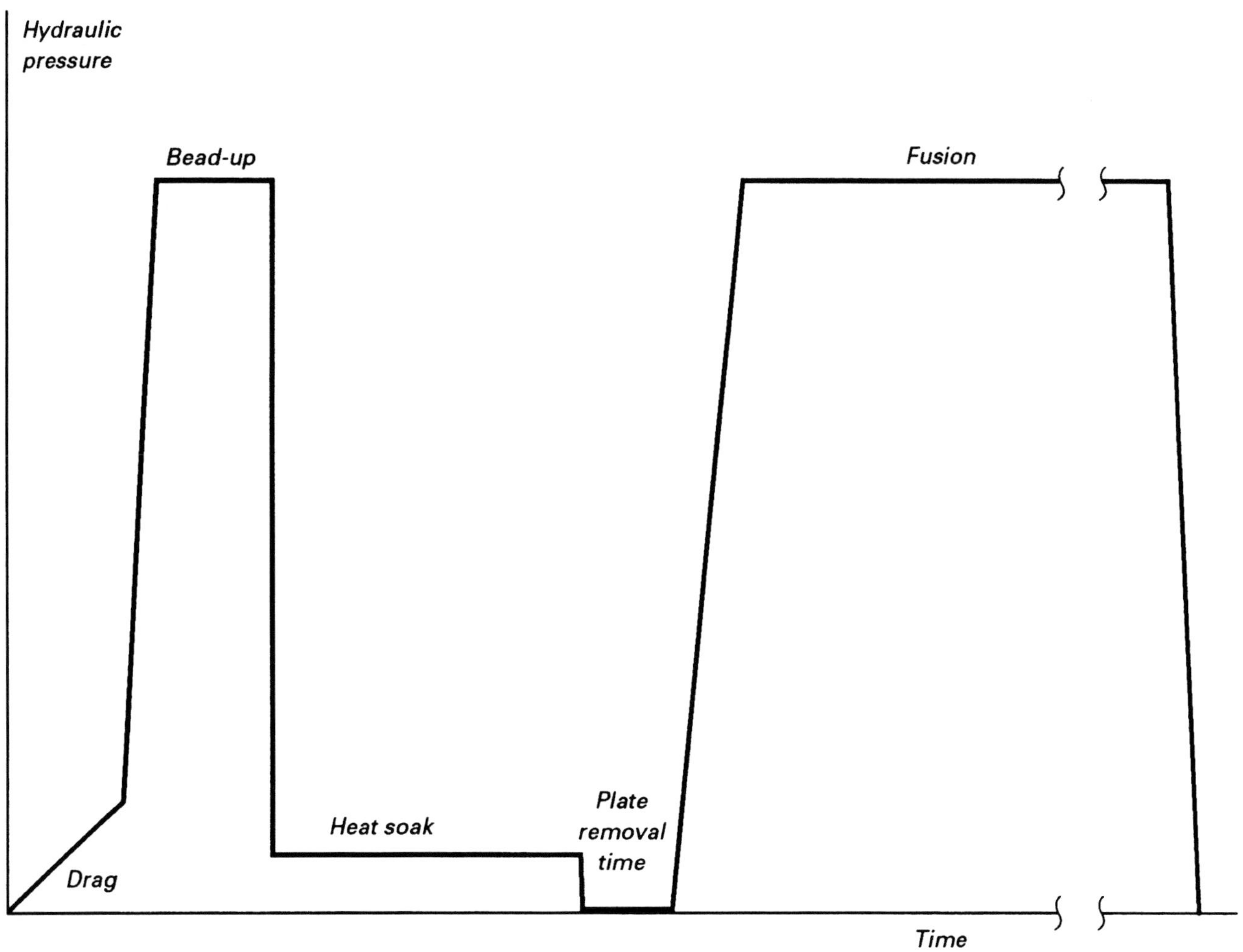

Fig. 4. Schematic diagram of the pressure cycle during the butt fusion welding process.

Fig. 5. Butt fusion welding equipment, showing the hot plate in position.

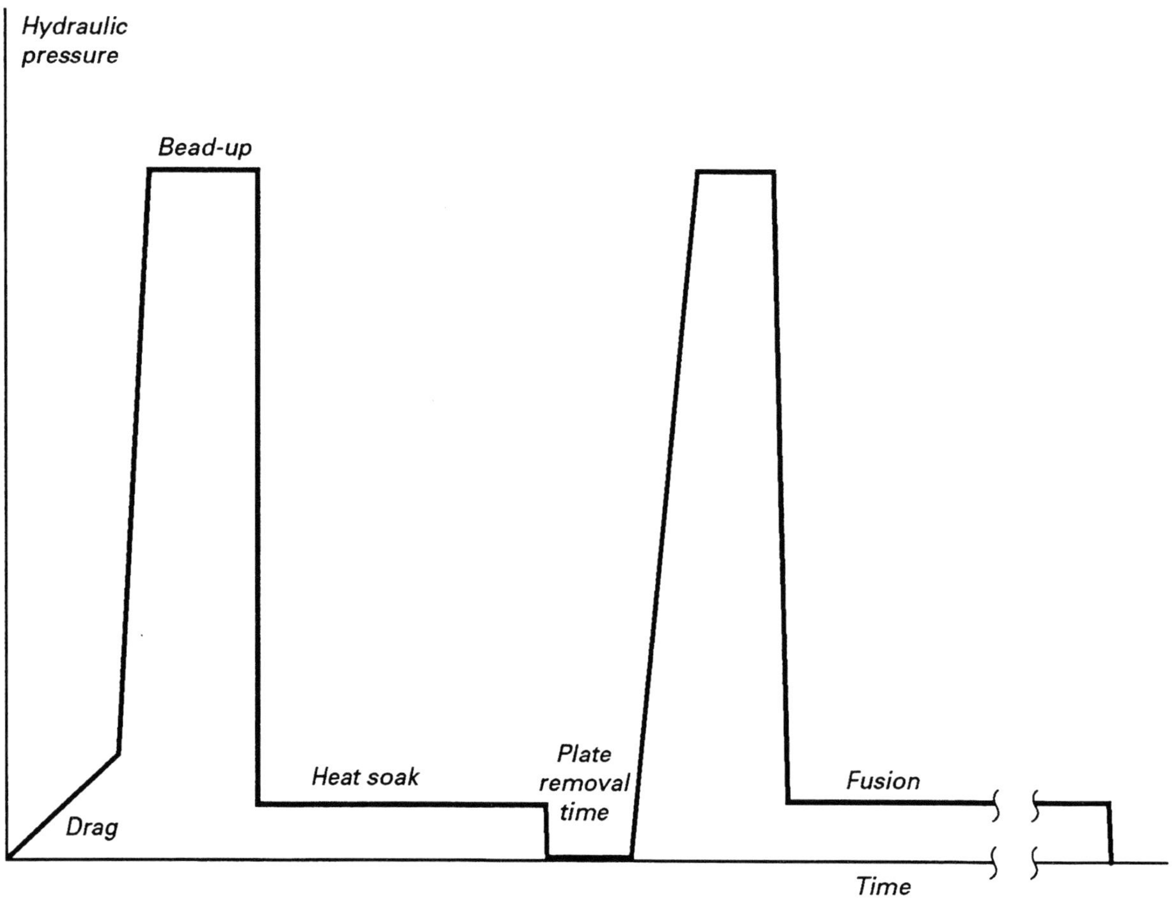

Fig. 6. Schematic illustration of the dual pressure butt fusion cycle.

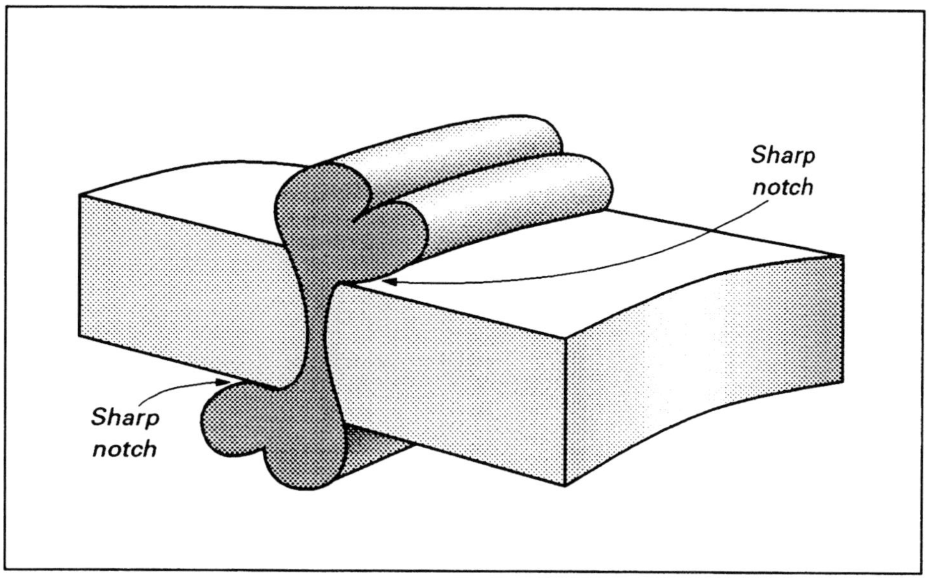

Fig. 7. Formation of weld bead due to pipe misalignment.

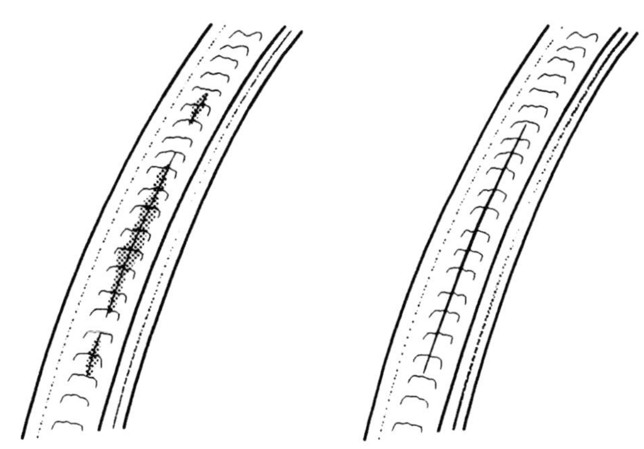

Fig. 8. Defects in pipe weld beads.

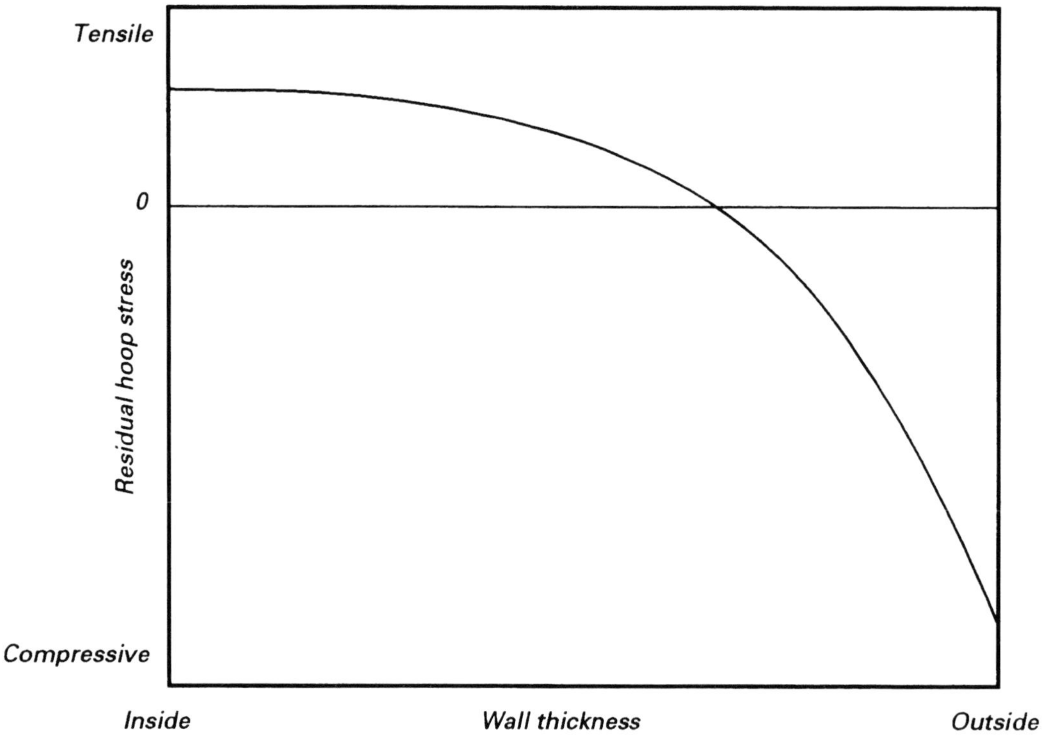

Fig. 9. Through-thickness variation of residual stress for externally cooled pipe, showing the existence of a tensile stress on the inside of the pipe wall and a compressive stress on the outside.

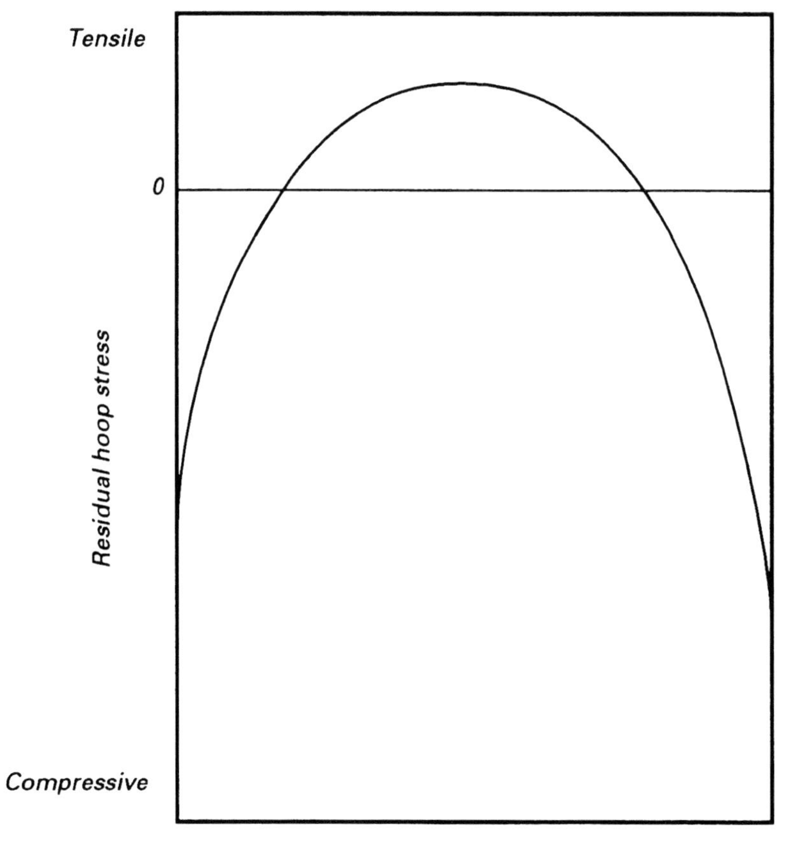

Fig. 10. Through-thickness variation of residual stress for pipe sumultaneously cooled on outside and inside, showing that the stresses are compressive on both the inside and the outside of the pipe.

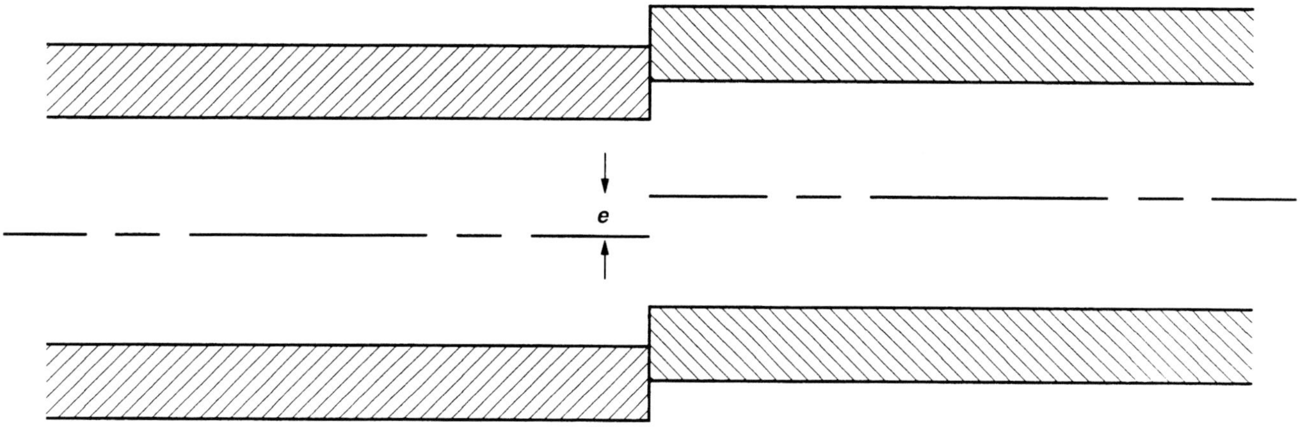

Fig. 11. Schematic illustration of radial displacement of a butt fusion welded pipe.

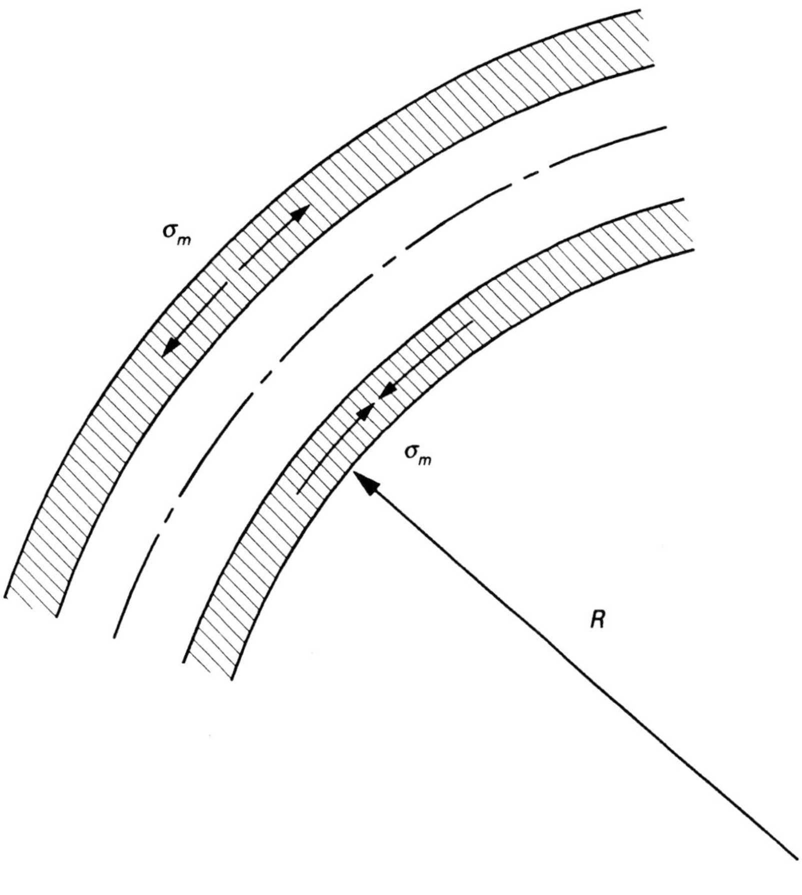

Fig. 12. Schematic diagram of axial stresses in a pipe wall due to bending.

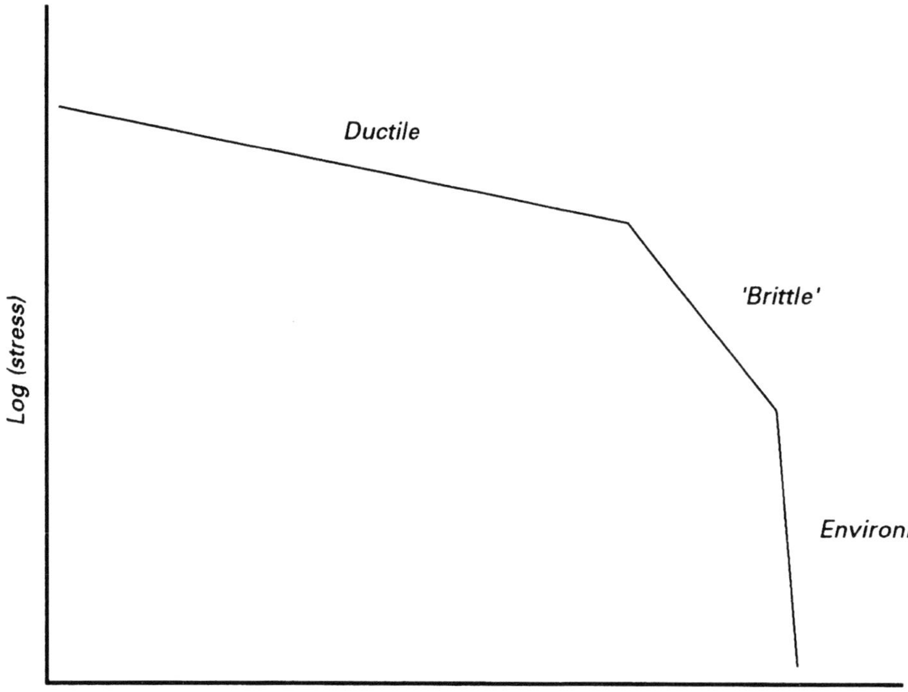

Fig. 13. Schematic stress rupture curve for a welded PE specimen.

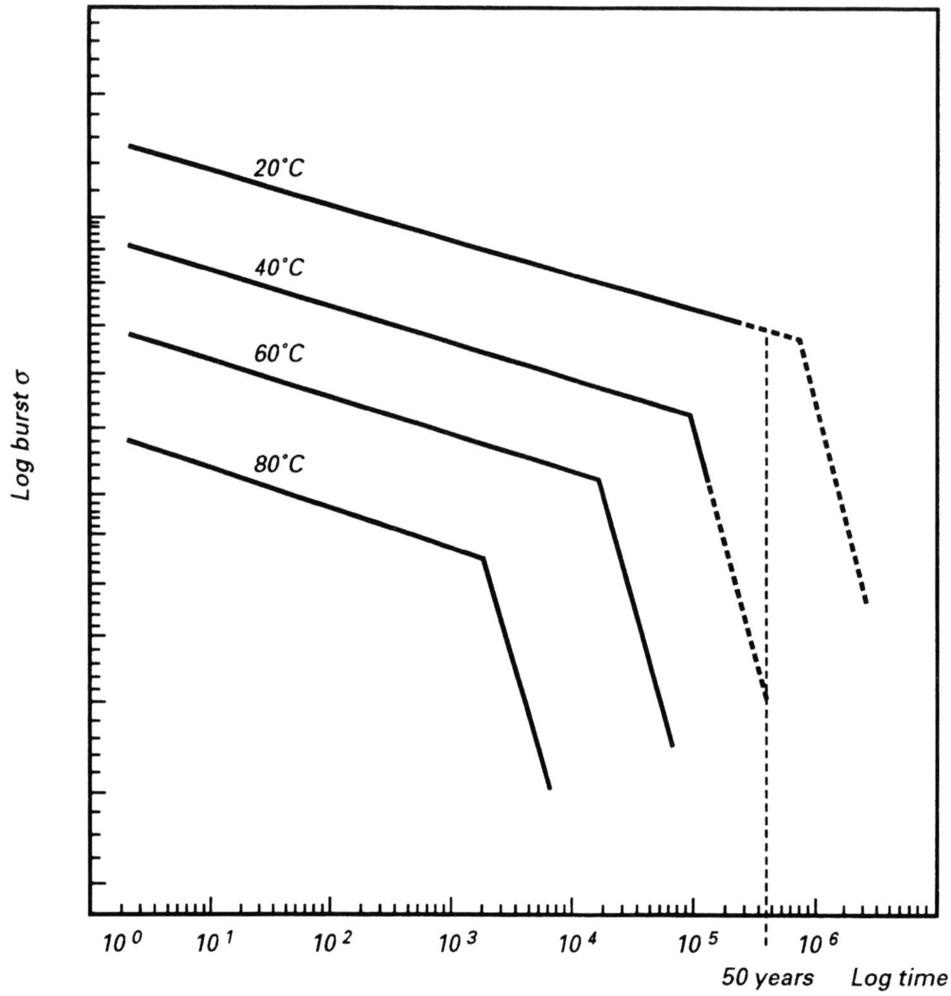

Fig. 14. Hydrostatic stress rupture curves for PE at various temperatures. The dotted lines are extrapolated. (2)

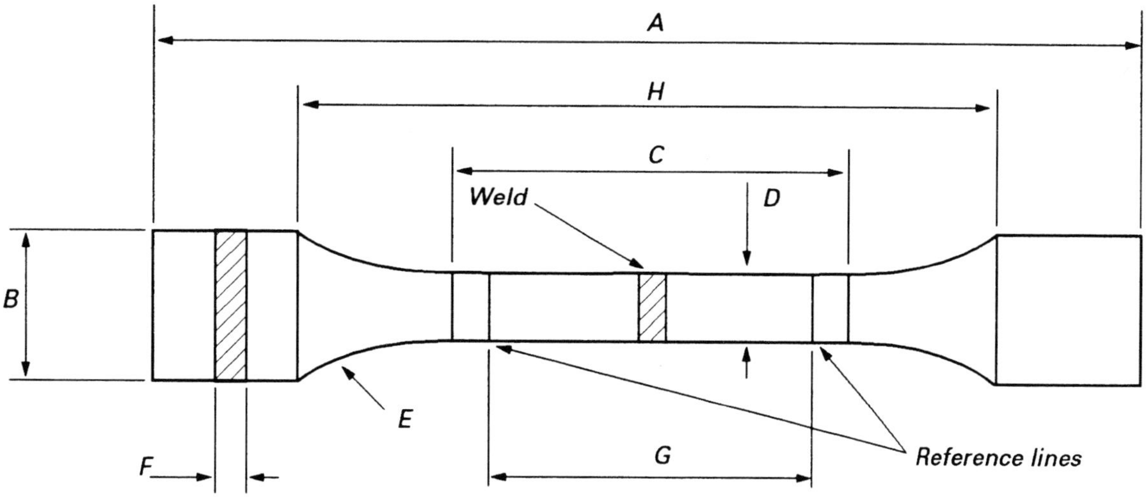

Dimensions are in millimetres		
A	Overall length, minimum	150
B	Width at ends	20 ± 0.5
C	Length of narrow parallel portion	60 ± 0.5
D	Width of narrow parallel portion	10 ± 0.5
E	Radius, minimum	60
F	Thickness minimum	1
	maximum	10
	preferred	3
G	Distance between reference lines	50 ± 0.5
H	Initial distance between grips	115 ± 5

Fig. 15. Tensile specimen geometry (91) adapted for welded samples.

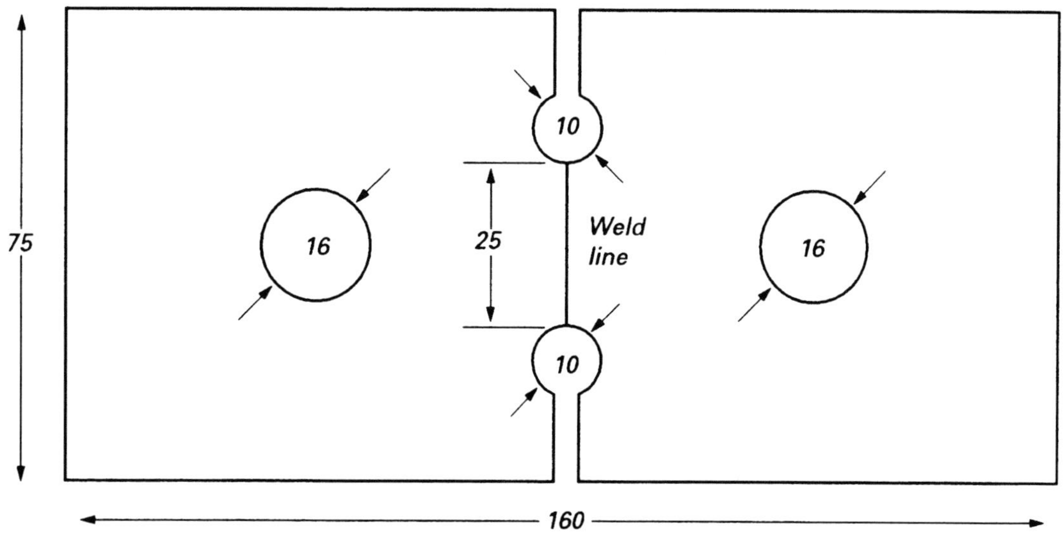

Fig. 16. Tensile test specimen geometry adopted for welded plastics pipe. (34)

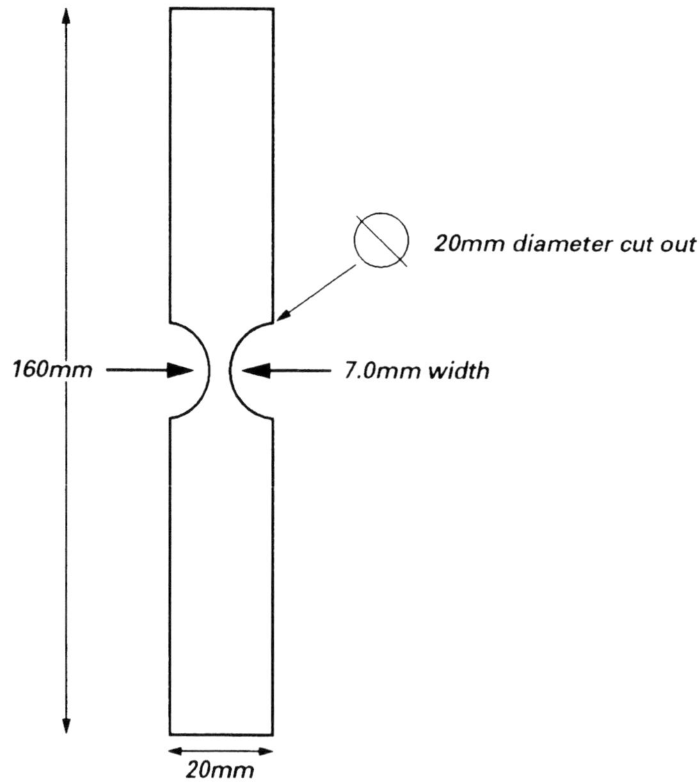

Fig. 17. Tensile specimen geometry developed by Georg Fischer. (86)

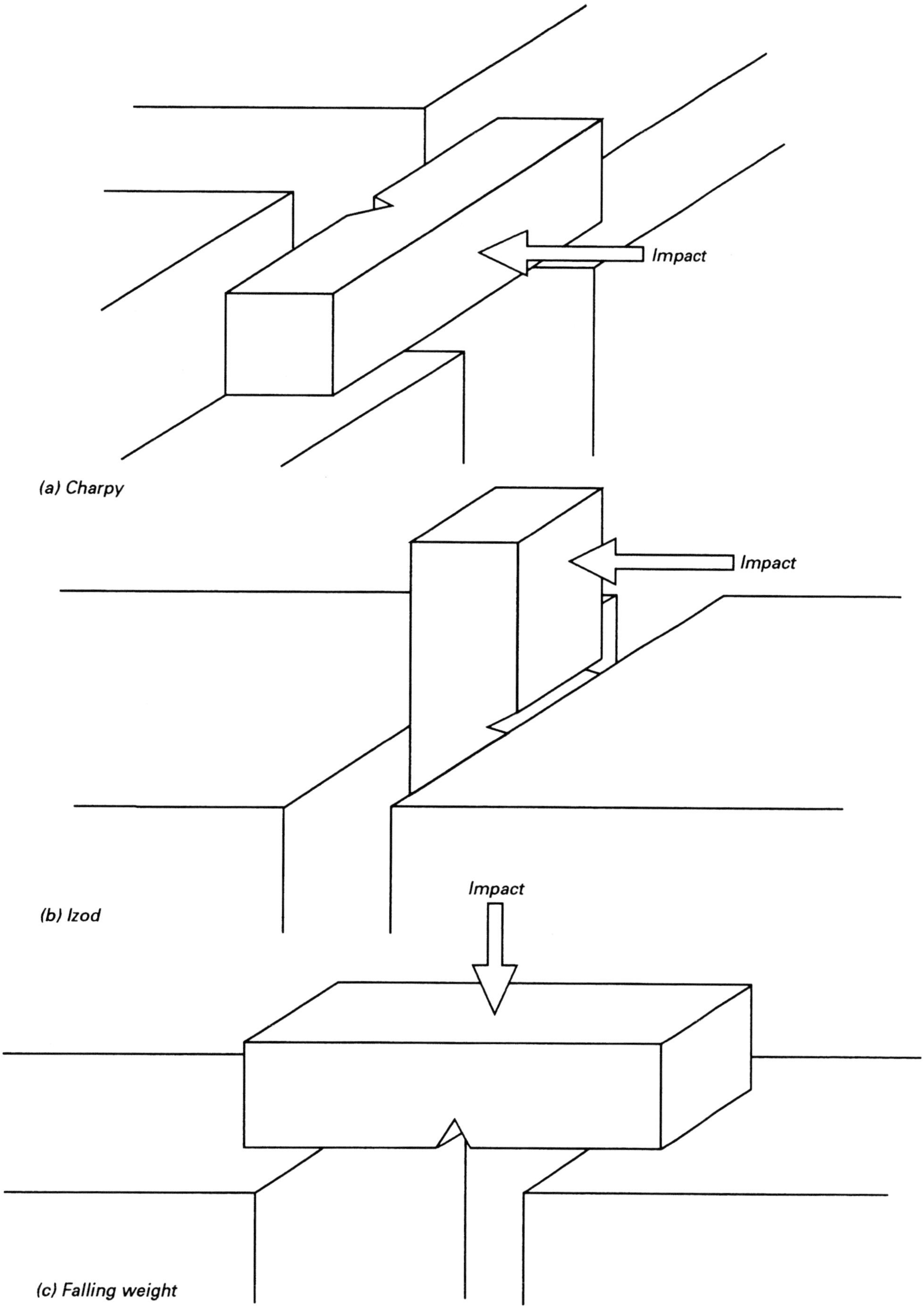

Fig. 18. Schematic illustration of various flexural impact tests.

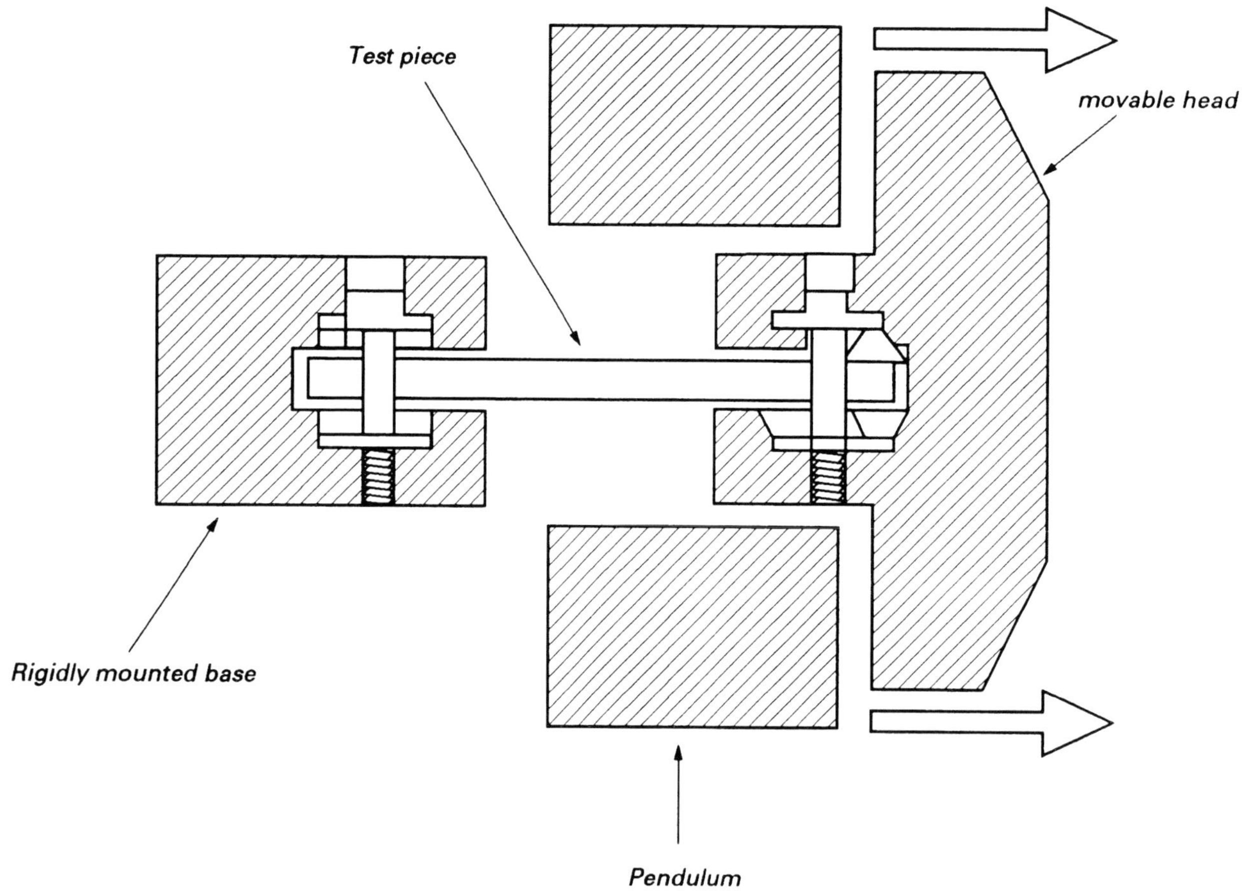

Fig. 19. Tensile impact test.

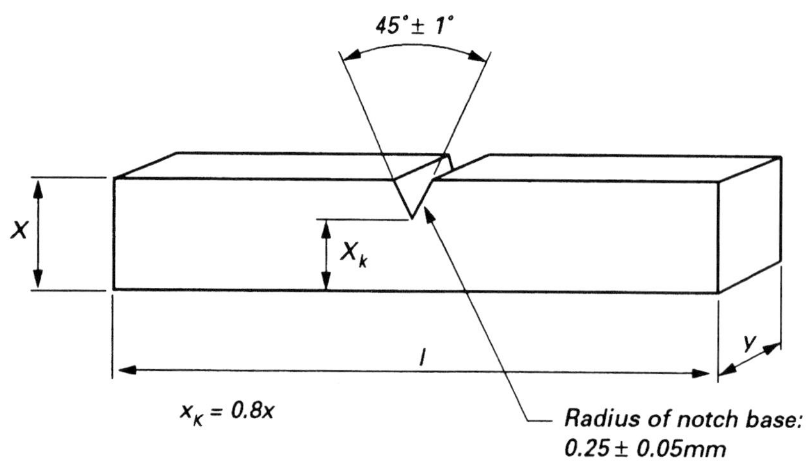

Specimen type	Length, l	Dimension, y	Dimension, x	Span between anvil supports	Remaining ligament, x_K
1	80 ± 2	10 ± 0.5	4 ± 0.2	60	3.2 ± 0.2
2	50 ± 1	6 ± 0.2	4 ± 0.2	40	3.2 ± 0.2
3	120 ± 2	15 ± 0.5	10 ± 0.5	70	8.0 ± 0.3

All dimensions in mm

Fig. 20. Charpy impact test specimen. (18)

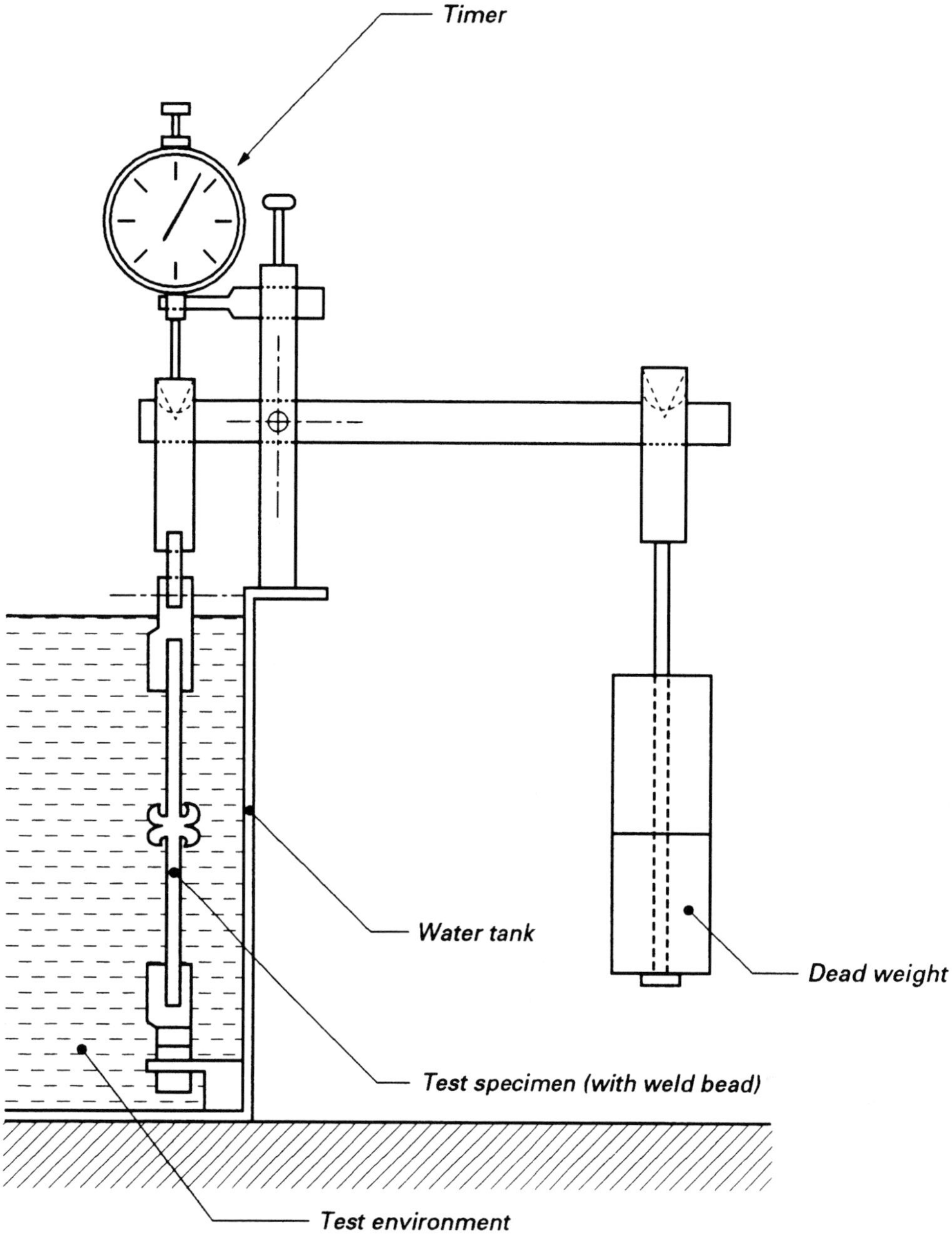

Fig. 21. Schematic diagram of the constant load test. (112)

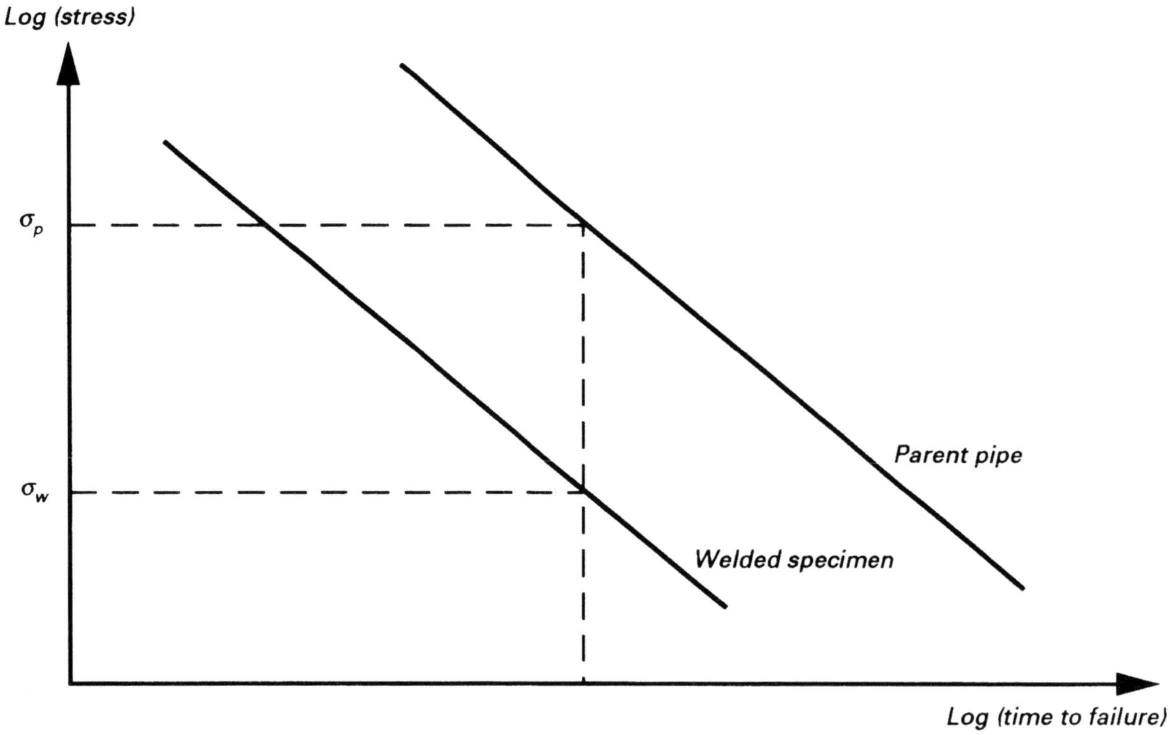

Long term tensile welding factor, $f_l = \frac{\sigma_w}{\sigma_p}$

Fig. 22. Schematic diagram of constant tensile load creep curves.

ULTRASONIC AND RADIOGRAPHIC NDT OF BUTT FUSION WELDED POLYETHYLENE PIPES

By: I.J. Munns and G.A. Georgiou

The use of polyethylene (PE) has increased in the water and gas industries over the past two decades. This article explains why, and highlights the need and development of reliable NDT techniques.

BACKGROUND

The popularity of plastic piping can be attributed to its lightness, flexibility, good corrosion resistance, and ease with which it can be joined. Increasingly, polyethylene is being used to replace metallic gas and water distribution pipelines, and is also being used for insertion repairs on leaking cast iron mains.

Polyethylene pipes with a diameter less than 250mm are normally joined using an electrofusion process, but for areas where the application is critical or the pipes are of larger diameter (>250mm) and of thicker section, the hot plate butt fusion process is preferred.

Quality of butt fusion joints in polyethylene pipe systems is primarily governed by control of the process parameters during welding. However, as new polyethylene materials are introduced (e.g. PE100) and increased demands are placed on existing polyethylene materials, there is an additional need to monitor quality through reliable non-destructive testing (NDT) methods. Currently there are no nationally accepted standards for the NDT of welds in polyethylene.

ULTRASONIC NDT

Previous work at TWI has shown that the attenuation of ultrasound in polyethylene is a factor of ten higher than for metals. In practical terms, this means that ultrasonic testing is limited to using low frequency compression waves (i.e. <4mHz) in order to achieve sufficient penetration.

An earlier study at TWI (11) considered the pulse–echo angled compression wave technique with polyethylene wedges to couple the ultrasonic energy from the transducer into the polyethylene specimen. Such ultrasonic procedures are largely insensitive to flaws which may exist in the vertical fusion face, such as inclusions, dust or 'cold' welds. One of the reasons for the poor performance of the pulse–echo technique is the weak response from vertically orientated flaws, and in a highly attenuative material like polyethylene. New work at TWI has considered pulse–echo creeping waves, the tandem and time of flight diffraction (TOFD) techniques, which are more suitable for detecting and sizing vertically orientated flaws.

Pulse-Echo Creeping Wave Technique

Creeping waves are high angle compression waves which are normally used on metals to detect surface or near surface cracks. Unfortunately creeping waves are only effective over a very short range. TWI have considered a pulse-echo creeping wave technique to inspect the fused area immediately beneath the outer weld bead. The sensitivity of the technique to lack of fusion and large material inclusions, located close to the pipe surface, has been demonstrated.

The Tandem Technique

The probe configuration for the tandem inspection technique is shown in Fig.8. By moving the transmitting probe (T) relative to the receiving probe (R) it is possible to obtain full through-thickness coverage of the welded region. The most appropriate arrangement for the tandem inspection of welds in 25mm thick polyethylene pipe was found to be using two 1MHz transducers, designed to generate compression waves at 60° in polyethylene.

Time-of-Flight Diffraction (TOFD) Technique

The probe configuration for the TOFD technique is shown in Fig.21 where two, 2.25MHz, 60° compression waves are used.

The first ultrasonic signal to be detected is the lateral wave, which travels the most direct route between transmitter and receiver. The final signal detected by the receiver is the reflection from the inner wall of the pipe, termed the backwall echo. The presence of a flaw will result in two diffracted signals, from the top and bottom edges of the flaw, being detected after the lateral wave and before the backwall echo.

The tandem and TOFD techniques were found to be sensitive to planar discontinuities, down to at least 2mm in diameter at the fused interface, but they are not sensitive to the presence of a cold weld. However, the presence of chalk dust at the fused interface causes a variation in the tandem and TOFD responses. Further work is needed to optimise these techniques for this flaw type.

X-RAY RADIOGRAPHIC NDT

In comparison to metals, plastics are much less absorbent of X-ray radiation and require the use of 'softer', lower energy, X-rays to achieve good quality results.

TWI have pioneered the use of wire type image quality indicators (IQIs) made from polyethylene to the specifications of BS EN462-1. This is a major achievement, and helps to overcome the problems of describing image quality of radiographs shot at low X-ray energy (<26kV).

The inspection procedures make use of the double-wall single image technique. When the X-ray source is in the plane of the weld the technique is referred to as the 'straight' technique, and the weld fusion face appears as a straight line on the radiograph. When the source is offset from the plane of the weld it is referred to as the 'throwing' technique and the weld fusion face appears as an ellipse on the radiograph. The straight technique was used as a primary method. The throwing technique was used to provide additional information regarding the shape and character of the flaw.

Prototype procedures have been used to detect reliably dust contamination of the fusion face, planar discontinuities in the fusion zone, lack of fusion and a thinning of the weld bead corresponding to the presence of a cold weld.

CONCLUSIONS

The conclusions are based on a range of polyethylene materials and pipes with diameters greater than 180mm and thicknesses greater than 12mm.

- Ultrasonic creeping waves can be used to inspect the fusion area immediately beneath the outer weld bead.

- Ultrasonic tandem and TOFD techniques can be used to detect and image perpendicular planar flaws.

- Ultrasonic techniques are currently unable to detect nominally cold welds.

- Radiographic intensities of between 16kV and 26kV are optimum when inspecting thicknesses between 5mm and 50mm.

- With radiography it is possible to infer the presence of a cold weld from local radiographic density measurements.

- Polyethylene wire type IQIs are optimum for IQI sensitivity, and have practical handling advantages with regard to pipe inspection.

ULTRASONIC AND RADIOGRAPHIC NDT OF BUTT FUSION WELDED POLYETHYLENE PIPES

By: I.J. Munns and G.A. Georgiou

1. INTRODUCTION

1.1. Background

The use of plastic pipe in the construction of fluid and gas distribution systems has increased dramatically over the past 30 years (1). The popularity of plastic piping can be attributed to its lightness, flexibility and good corrosion resistance, as well as the ease with which it can be joined. Pipes with a diameter <250mm are normally joined using the electrofusion process. However, for areas where the application is critical, or the pipes are of larger diameter (>250mm) and of thicker section, the butt fusion process is preferred. Both of these heat fusion processes are capable of producing a joint with mechanical properties approximately equivalent to those of the parent material (2).

The butt fusion process is described in detail in earlier work, (3,4,5) but a brief description is given here. The process involves bringing the prepared end surfaces of two lengths of pipe in contact with a heated plate. After a specified time, the hot plate is removed and the molten pipe ends are then pushed together and held under pressure until they have cooled sufficiently to allow handling. The butt fusion process produces 'weld beads' on the inner and outer pipewalls. The time between removing the plate and then pushing the ends together is the 'plate removal time' or sometimes referred to as the 'delay time'.

British Gas (6) have developed design methods for butt fusion welded polyethylene pipeline systems based on a minimum service life of 50 years at a temperature of 20°C. It is recognised, however, that the stress analysis and fracture mechanics techniques, used to determine the operational lifetime of a polyethylene pipe system, requires certain assumptions to be made about the pipe material and the way in which it is loaded and, because of this, it is necessary to incorporate a level of conservatism into the design of polyethylene pipe systems (7).

Despite conservative design rules, in practice, the overall lifetime of a piping system is entirely dependent on correctly specified materials and construction standards, and the effectiveness of a quality control procedure designed to ensure that these specifications and standards are met on a day to day basis. Failure to meet these requirements can result in a defective joint. For example, if the pipe ends are not adequately cleaned prior to welding, there is the possibility of contaminants such as dust, sand, soil, water, oil etc. being introduced at the weld interface (7). Also, if, during the butt welding process, the hot plate temperature is too low, or the delay time between removing the hot plate and joining the pipe ends is too great, the surfaces of the pipe ends will crystallise before they come into contact and a 'cold' weld will be formed. For 25mm thick medium density polyethylene pipe, a nominally cold weld is associated with a plate removal time of the order of 40 seconds and above (4).

Failure of a butt joint, as a consequence of these defects, is likely to result in an open end fracture of the pipe rather than joint leakage. In a gas distribution pipeline, such a failure presents a serious hazard.

One method of assuring quality on-site is to subject the pipeline system to a 24 hour proof test at 1.5 times maximum operating pressure. Even this, however, fails to verify completely the long term integrity of the system after construction. Ewing and Richard (6) describe a situation where a pipe with an 80% wall thickness defect would pass a 6 bar pressure test, yet the same pipe would show a reduction in 50 year design life of some 53%. From this it is clear that other methods of assuring long term integrity must be sought.

In addition to operator training and the enforcement of stringent quality control procedures, the development of suitable methods of non-destructive testing (NDT) have a significant role to play in ensuring the overall integrity of butt fusion welded polyethylene piping systems (6,8).

At present there are no nationally accepted standards for the NDT of welds in polyethylene. The most widely used technique is a visual inspection of the outer weld bead. This method is only sensitive to gross flaws such as pipe misalignment or defects inferred from the presence of an abnormal weld bead. The limitations of a visual technique, such as this, must be realised; a poorly formed bead may reliably indicate an unsatisfactory weld, but a correctly formed bead is not always a sure indication that the weld is satisfactory and free from internal defects.

Attempts have been made to use more quantitative methods of weld examination. X-ray radiography is routinely used for the inspection of large diameter (>300mm) polyethylene pipe (9), but problems exist in describing the image quality of the radiographs (6). Also, at one stage a guideline standard (ASTM F600-78) for the manual ultrasonic inspection of butt fusion welds in polyethylene pipe was introduced, but the results obtained using this standard were so heavily dependent upon the skill of the operator that it was withdrawn in 1991.

To summarise, at the present time, it is routine to inspect the quality of butt fusion welds in polyethylene pipe using a visual examination of the outer weld bead and, in a small number of instances where larger diameter pipe is to be inspected, to use radiography. Both of these techniques are known to have practical limitations. For example, a defect in thick wall pipe may not manifest itself in the weld bead, and radiography is only applied to a small percent of the welds. Therefore, there is a need for additional work to improve the reliability of existing methods of NDT, in addition to developing other methods of inspection.

1.2. Previous Work at TWI

TWI have carried out much work in the development of ultrasonic and radiographic techniques for the inspection of butt fusion welded polyethylene pipe (10,11). In (11) the authors studied the effect of material composition on ultrasonic wave velocity and attenuation. Separate studies were also included to assess the degree of anisotropy (i.e. differences in elastic properties in different directions) present in the polyethylene materials considered. It was concluded that ultrasonic wave velocity and attenuation do not vary significantly with material composition and that, when inspected ultrasonically, polyethylene exhibits very little anisotropy and no adjustment in sensitivity is required when testing in different directions. This earlier study also reports on the effectiveness of conventional ultrasonic pulse–echo techniques for detecting a range of natural and manufactured defects in sections of 25mm thick polyethylene pipe, joined using the butt fusion process. The sensitivity of the technique to a range of defects, including inclusions (e.g. material inserts and dust) and cold welds was established. The pulse–echo techniques evaluated were not capable of detecting any of the above flaws with any reliability. One of the reasons for the poor performance of the pulse–echo technique is the weak response from vertically orientated flaws in a highly attenuative material like polyethylene. This was highlighted in (11) using three 10mm diameter flat bottomed holes drilled at different through–thickness positions at one of a pipe section, and is illustrated here in Fig.1. Each hole was detected by strong reflections from the corner between the hole and the end of the pipe, rather than from specular reflections from the face of each hole. The earlier work did, however, identify other ultrasonic techniques, such as creeping waves, time–of–flight diffraction (TOFD) and the tandem technique, as being potentially more suited to the inspection of butt fusion welded polyethylene pipe.

With respect to radiographic inspection, TWI have pioneered the use of wire type image quality indicators (IQIs) made from polyethylene to the specifications of BS EN 462–1 (12). The existence of such IQIs is a major achievement and significantly helps to overcome the problems of describing image quality of radiographs shot at low X–ray energy's. The same programme of work (11) considered the effect of X–ray absorption for different grades of polyethylene. It was concluded that the very low absorption by polyethylene of X–ray radiation can vary significantly from one polyethylene type to another but that, for the range of polyethylene materials shown in Table 1, only two exposure charts are necessary to produce radiographs in the density range specified by BS 2600 (13). These exposure charts formed the basis of radiographic procedures which were then used to evaluate the same specimens considered by the ultrasonic pulse–echo technique. Radiography proved far more successful than the pulse–echo ultrasonic technique, and the radiographic procedures were able to locate manufactured flaws such as foil inserts and chalk dust within the fusion zone and, more significantly, the presence of a cold weld (11).

This report describes further research into NDT of polyethylene. The ultrasonic procedures have been developed beyond the basic pulse–echo approach and the detection performance and sizing capabilities have improved significantly. New work on ultrasonic techniques such as creeping waves, tandem and TOFD are reported in Section 3. Specific aspects of the need and implications of using low kV radiography for testing polyethylene and the development of a standard for polyethylene IQIs is considered in Section 4. Further details regarding the detection and significance of cold welds is also given in this section. The results of this study are discussed in Section 5 and the main conclusions are given in Section 6. The latest ultrasonic and radiographic procedures for testing butt fusion welds in polyethylene pipe are given in Appendices A and B respectively.

2. OBJECTIVE

The overall objective of this work has been to improve the quality control of butt fusion joints in polyethylene pipes through the availability and application of newly developed NDT procedures.

3. ULTRASONIC NDT

3.1. Background

There are fundamental differences between the ultrasonic inspection of polyethylene and the ultrasonic inspection of metals. Previous work at TWI (11) has shown that the attenuation of ultrasound in polyethylene is a factor of ten higher than for metals. Figure 2 shows that this attenuation is directly proportional to the frequency of vibration of the ultrasound, and also shows that polyethylene attenuates ultrasonic shear waves much more rapidly than compression waves. Essentially, this means that for many practical applications, ultrasonic testing is limited to using low frequency compression waves (i.e. <4MHz) in order to achieve sufficient penetration and sensitivity on typical thicknesses (25mm) and grades of polyethylene. This section will concentrate on methods of ultrasonic testing using creeping waves, tandem and TOFD techniques. Each of these techniques has been evaluated on a range of polyethylene pipe specimens containing precise defects.

3.2. Defect Specimens

Throughout this study five polyethylene pipe specimens were evaluated (see Table 2.) Each specimen consisted of a butt fusion weld between two sections of Finathene 3802 Y (yellow) polyethylene pipe, 250mm external diameter and 25mm thick, typical of that used by the UK gas distribution industry. All five specimens were manufactured at TWI and two of these specimens, designated H59–1 and H59–3, were welded under optimum or standard conditions (5). The other three specimens were fabricated to simulate a range of flaws representative of those potentially arising during the routine installation of a butt fusion welded polyethylene piping system. Specimen H59–5, manufactured to simulate particle contamination of the fusion face (dust, soil etc), had chalk dust added to one half of a molten pipe end prior to fusion (Fig.3(a)). Another, H59–4, simulating a weld containing isolated areas of lack of fusion, had a range of different diameter aluminium foil discs (2, 4, 6, 8 and 10mm) attached to its interface prior to welding (Fig.3(b)) and one, H59–2, had a 50 second delay time from removal of the hot plate to joining of the pipe sections (i.e. nominally a cold weld).

3.3. The Creeping Wave Technique

If the angle between a source of compression waves and the surface of a material is such that incident compression waves are refracted to travel parallel to this surface, then this angle is termed the first critical angle (Fig.4). Shortly before this critical angle is reached, another wave is observed which propagates parallel to and just beneath the material's surface (Fig.5), and is termed a 'creeping' wave (14). In such cases, a mode converted shear wave is still transmitted into the material. As a creeping wave propagates, it interacts with the surface of the material and generates a series of secondary shear waves. The effects of beam divergence mean that, in addition to the creeping wave other high angle compression waves are also generated, typically at angles of 70° and above.

For the inspection of steel, creeping wave probes are commercially available and are used to detect surface breaking cracks and near surface defects. Unfortunately, the effectiveness of creeping wave probes is limited to a relatively short range, as energy is continuously being converted into secondary shear waves. This attenuation problem is particularly apparent when generating creeping waves in polyethylene, and it becomes necessary to maximise the strength of the ultrasound signal within the polyethylene through specialised probe design. Careful selection of the probe wedge material can be used as means of improving the intensity of ultrasound generated within the polyethylene.

TWI considered polytetrafluoroethylene (PTFE) as a creeping wave wedge material for use on polyethylene. Notwithstanding its high attenuation (755dB/m at 2.25MHz), PTFE gives reasonable acoustic coupling to polyethylene but, most importantly, its compression wave velocity is much less than that of polyethylene (i.e. 1380m/sec as opposed to 2300m/sec). This enables only a small inclination of the PTFE wedge to the surface of the polyethylene to produce the necessary refraction of the ultrasonic beam. Consequently, the size of the probe shoe need not be large and attenuation of ultrasound through the wedge can be minimised, thus maximising the strength of the creeping wave generated in the polyethylene. To generate creeping waves in polyethylene a PTFE wedge angle of only 36° is required (Fig.5).

Pulse–echo experiments on a 10mm thick polyethylene test block containing very fine (<250μm) saw–cuts were used to determine the sensitivity of the creeping wave technique, and to assess its suitability as a method for inspecting the region immediately beneath the outer weld bead on polyethylene pipe. During the experiments it became apparent that the generation of creeping waves in the polyethylene block was heavily dependent on the quality of coupling at the probe/specimen interface. For this reason it was necessary to use a large amount of water–based couplant during scanning.

Three creeping wave probes with centre frequencies of 1, 2.25 and 3.5MHz were evaluated. Each probe was positioned 5mm away from the free edge of the test block and the magnitude of reflected creeping wave signals were compared. The 1MHz probe gave a response that was well over 20dB higher than an otherwise identical 3.5MHz probe. At a range of 5mm, both the 1MHz and 2.25MHz frequency probes could detect a saw–cut extending to a depth 3mm from the block surface. However, the magnitude of the reflected creeping wave signal decays rapidly as the distance between the probe and saw–cut is increased (Fig.6). Experiments have been carried out using other, less attenuative, wedge materials in an attempt to increase the strength of the creeping wave generated in the polyethylene. Both polybutadiene (attenuation of 420dB/m at 2.25MHz) and natural rubber (attenuation of 480dB/m at 2.25MHz) probe shoes have been evaluated, but no significant improvement in the range of creeping wave propagation has been attained.

A 1MHz pulse–echo creeping wave scan was performed on pipe specimens H59–1 to H59–5 inclusive. The creeping wave probe was positioned against the outer weld bead and then scanned around the circumference of each pipe. During scanning, a strong reflection from the outer weld bead is detected and is clearly identifiable on the A–scan display. This signal can be used to ensure that transmission of ultrasonic energy from the ultrasonic transducer into the polyethylene pipe is maintained during scanning.

In the inspection of H59–4, the creeping wave A–scan response from the 6mm diameter foil insert is shown in Fig.7. During fabrication of this specimen each foil disc was positioned in the mid–thickness of the pipewall, and an inspection using creeping waves, which are only sensitive to imperfections close to the surface where they are generated, would have not been expected to detect very much. Clearly, in this case, the final position of the 6mm disc is such that interaction with the creeping wave does occur. For specimen H59–4, the creeping wave technique was able to detect the presence of both the 6 and 10mm diameter discs, but not the other discs. A creeping wave inspection of the two other defective specimens, H59–5 (chalk dust) and H59–2 (cold weld), did not detect any significant variation in response from the fusion area immediately beneath the outer weld bead.

Creeping waves have demonstrated a sensitivity to gross discontinuities, such as fine saw–cuts and aluminium foil inserts, located close to the surface where creeping waves are generated. They have not, however, been able to detect chalk dust at the fused interface or the presence of a nominal cold weld.

3.4. The Tandem Technique

The probe configuration for the tandem inspection technique is shown in Fig.8. On metals, tandem inspection is most commonly applied to welded joints in thick walled vessels. Two angled probes, designed to generate 45° shear waves in the material, are used and, by adjusting the probe separation distance, it is possible to obtain full coverage of a considerable thickness of weld. Tandem inspection is particularly sensitive to flaws orientated perpendicular to the scanning surface and, for this reason, is potentially suited to detecting lack of fusion type defects in butt welded polyethylene pipework. In this application, however, it is only practical to generate angled compression waves in the polyethylene, as shear waves are attenuated too rapidly. In addition, for typical pipewall thicknesses of around 25mm, the separation distance between the transmitting and receiving transducers is significantly reduced, and the use of 45° probes becomes impractical. This is the case when considering the tandem technique to inspect the 25mm thick polyethylene pipe specimens used in this study. To overcome this problem, all tandem experiments discussed in this report were performed using PTFE wedges (i.e. wedge angle is 30°), and compression wave transducers in order to generate 60° angled compression waves in the polyethylene. The base of each wedge was modified to fit the curvature of the pipe.

Preliminary experiments showed that, for the tandem inspection of 25mm thick polyethylene, 1MHz probes gave an improved signal-to-noise performance over 2.25MHz probes. A probe frequency of 1MHz was selected for all subsequent tandem scans. Figure 9 shows a typical A-scan response from a perpendicular reflector detected using the tandem technique. By using TWI's P-scan PSP-3 digital ultrasonic flaw detector to collect such A-scans as the probes are scanned around the circumference of the pipe, it is possible to process these data and to display the information as a composite A-scan image. Figure 10 shows a composite A-scan image of a section of pipe containing five end-drilled holes. Each hole is 5mm in diameter, has a flat bottom and extends 20mm in from the free end of the pipe. All five of the holes are clearly visible on the composite A-scan image. The signal on the right hand side corresponds to the reflection off the end of the pipe itself. Throughout the scan, the transmitting and receiving probes were kept a fixed distance apart as, in this thickness of material, it is not practical to move the transmitting and receiving probes relative to each other in order to achieve full scanning coverage of the pipe thickness. At this particular probe separation distance the probes are more sensitive to perpendicular flaws in the mid-thickness region of the pipewall (maximum signal response). However, the beam of ultrasound entering the pipe is sufficiently divergent to enable a small amplitude response to be received from the holes nearest the top and bottom surfaces of the pipewall. Therefore, it is possible to obtain through-thickness coverage of a 25mm thick weld, without the need to adjust the separation of the transducers during tandem inspections.

The tandem technique was then applied to pipe H59-4, containing aluminium foil inserts at the fused interface. The entire circumference of this pipe was scanned using the P-scan system. Figures 11-14 are examples of the composite A-scan images recorded. As well as a strong amplitude response from the foil insert, an additional echo was detected. This appeared as a constant amplitude signal at a particular distance (range) along the A-scan trace. Range calculations have indicated that this echo originated from the inner weld bead and studies on a pipe with the inner bead removed have verified this. The size and position of this signal was not affected, in anyway, by the presence of a defect and, for this reason, the signal could be used as a check to see whether consistent coupling of the probes to the polyethylene pipe surface was being achieved during scanning.

The tandem technique was able to detect reliably the 2, 4, 6 and 10mm diameter foil inserts in pipe H59-4. However, during scanning of this defect specimen, it was not possible to detect the 8mm diameter foil disc. An analysis of the maximum amplitude response from the other discs showed that there was little correlation between the size of the reflected signal and the size of the disc insert (Fig.15). It is believed that this was because the position of the foil inserts in the pipe thickness had altered during the welding process. Prior to fusion, each aluminium foil disc was originally centred on the mid–thickness position of the pipewall. At the fusion stage, when pressure is applied to force the molten ends of each pipe section into contact, it is possible that some of the discs may have been dislodged from the mid–thickness position to a position closer to the inner or outer weld beads. In these extreme positions a disc would give a smaller signal amplitude response during tandem inspection than if it were positioned in the mid–thickness of the pipewall.

The section of pipe containing the 8mm diameter foil insert was removed and radiographed to reveal the position of the disc in the through–thickness position (Fig.16). For comparative purposes, a similar section containing the 10mm foil disc insert was also removed and radiographed (Fig.17). It is evident that the 8mm diameter foil disc is positioned slightly off–centre. This displacement could be sufficient to cause a reduction in amplitude of the reflected signal, however, it is certainly not sufficient for the disc to go undetected by the tandem technique.

Additional normal incidence pulse–echo experiments, using a 2.25MHz compression probe on the sections containing the 8 and 10mm diameter discs, showed that the signal reflected from the 8mm diameter disc was much lower in amplitude than anticipated. In fact, this signal was over 20dB (i.e. a factor of 10) below the response recorded from the section containing the 10mm diameter disc. In the same experiments the ultrasound transmitted through the foil disc and reflected off the backwall of the specimen was monitored (Fig.18). For the 10mm diameter disc, the strong reflected signal from the disc itself was accompanied by an almost total loss in signal from the back wall. In contrast, the A–scan response from the 8mm diameter foil insert was dominated by a strong backwall reflection, which changed little in amplitude as the probe was moved off the area containing the disc on to the parent material.

Clearly the 8mm diameter foil insert, unlike the 10mm diameter foil insert, is not behaving as a smooth planar reflector. In fact, the 8mm diameter foil disc appears to be partially transparent to ultrasound. At the present time, the reasons for this are unconfirmed. However, a plausible explanation of this behaviour is that, during the fusion process, molten polyethylene material did not adhere to the 2, 4, 6 and 10mm diameter foil discs and each of these discs is now surrounded by a very thin layer of air, whereas, during the same welding process, molten polyethylene material did adhere to the 8mm diameter foil disc forming a well bonded interface between polyethylene and aluminium. As a result, during ultrasonic inspection, the 10mm diameter and 8mm diameter foil discs exhibit distinctly different reflection characteristics. The amount of ultrasonic energy transmitted across or reflected from an interface is dependent on the relative acoustic impedance of the two interface materials. Acoustic impedance is the product of density and ultrasonic wave velocity of the medium. Figure 19 shows schematically what is believed to be the nature of the interface between the polyethylene and the 8mm diameter foil disc and the polyethylene and 10mm diameter foil disc. The reflection coefficient for each case is calculated, and it is shown that a much stronger response from the 10mm diameter foil disc would be expected under the assumed circumstances, described above. The combination of off–centre position and altered reflection characteristics for the 8mm diameter foil insert, are believed to account for why this insert remained undetected by the tandem technique.

A tandem scan of pipe H59–5 showed a variation in signal received from areas known to contain chalk dust and those that are dust free (Fig.20). At the present time, it is unclear whether this variation is solely due to the presence of chalk dust or whether it is due to another feature of the weld, such as a variation in bead condition. Further work is needed to establish the sensitivity of the tandem technique to chalk dust at the fusion face.

Separate tandem scans on specimens H59–2 (50 second delay) and H59–3 (standard weld) failed to discriminate between a standard weld and a weld fabricated with an extended delay time, that is, a nominally cold weld.

To summarise, the tandem technique has demonstrated an ability to detect planar discontinuities, down to 2mm in through–thickness height. Also, the sensitivity of the technique to dust contamination of the fusion face has yet to be confirmed, and such a technique is unable to detect the presence of a nominally cold weld.

3.5. The Time–of–Flight Diffraction (TOFD) Technique

In weld inspection of metals, TOFD is often used to determine the through–thickness height of flaws. The technique is especially sensitive to flaws orientated perpendicular to the scanning surface and, for this reason, is considered in this study of butt welds in polyethylene pipe.

The probe configuration for TOFD is shown in Fig.21. In the following experiments, two 2.25MHz 60° compression wave probes were used. During TOFD inspection, ultrasound is generated in a divergent beam which isonifies the complete thickness of the fusion zone. Displayed on a time–trace, the amplitude signals detected by the receiving probe would look similar to Fig.22. Each signal is separated in time according to the distance ultrasound has to travel between the transmitting and receiving transducers. The first ultrasonic signal to be detected is the lateral wave, which travels the most direct route between transmitter and receiver. The final signal detected by the receiver is the 'bounce' off the inner wall of the pipe, and is termed the 'backwall echo' (Fig.21).

A perpendicular defect in the weld line will diffract ultrasound from its top and bottom edges and some of this diffracted energy will be detected by the receiving transducer. Diffracted signals appear as low amplitude responses, between the lateral and backwall echo signals on the time trace (Fig.22). In TOFD inspection of metals, it is possible to accurately measure the through–thickness height of a flaw by scaling the separation between the diffracted signals. However, the lower frequencies and longer wavelengths required for polyethylene inspection mean that it is not always possible to identify the signals diffracted from the top and bottom of the flaw, and accurate sizing of flaws in polyethylene is generally problematic. A–scans, collected by the tandem technique, are commonly displayed as TOFD D–scan images of the fusion area.

The following TOFD experiments were conducted using a pre-amplifier attached to the receiving transducer, as this improves the signal-to-noise ratio by about 30dB (i.e. a factor of 30.) TOFD scans from specimen H59-4, containing aluminium foil inserts at the fused interface, are shown in Fig.23-26. The presence of each disc is clearly discernable on the images produced. Again, however, there is no strong response from the 8mm diameter foil disc.

The TOFD technique relies on the presence of a discontinuity in the path of incident ultrasound to diffract ultrasound from its edges in order for it to be detected. For this reason it is doubtful whether TOFD inspection would be sensitive to a nominally cold weld in polyethylene. A nominally cold weld is synonymous with poor fusion across the entire welded interface and there is no local discontinuity from which diffracted signals can originate. This was verified by scanning specimen H59-2 (50 second delay) and showing that the TOFD response was no different to that received from pipe H59-3 (welded under optimum conditions).

A TOFD scan of pipe H59-5 (chalk dust) appears to show a subtle variation between an area of the weld containing chalk dust and an area that is dust free (Fig.27). However, as for the tandem inspection of this specimen, further work is required before the sensitivity of TOFD to dust contamination of the fusion face can be established.

In summary, the TOFD technique has detected reliably gross planar discontinuities, down to 2mm in through-thickness height, at the fused interface. The sensitivity of the TOFD technique to dust contamination of the fusion face has yet to be confirmed and as expected, TOFD was not able to detect the presence of a nominally cold weld.

3.6. Updated Procedures for the Ultrasonic Inspection of Polyethylene Pipes

The draft ultrasonic procedures included in Appendix A of this report could potentially form the basis of a standard. Currently, a standard detailing the ultrasonic examination of butt fusion welds in polyethylene pipes does not exist. However, future work at TWI is planned in order to develop and extend the procedures in Appendix A to form a document for submission to the relevant BSI committee. The introduction of such a standard will represent a significant improvement in the quality assurance for butt fusion welds in polyethylene pipes.

4. X-RAY RADIOGRAPHY

Following the publication of the earlier TWI Members report (11), technical discussions with Member companies raised a number of important issues, mainly in connection with the prototype radiographic procedures outlined in that report. These and other technical issues are dealt with in this section and, as a result, the radiographic procedures for examining fusion welded circumferential butt joints in polyethylene pipe are updated and presented here in Appendix B. The work on radiography is at a stage where a draft radiographic standard can be submitted to the relevant BSI committee for consideration.

4.1. The Implications of Using Low kV

The exposure charts given in (11), for each of the polyethylene materials considered, use radiographic intensities between 10kV and 16kV inclusive. At such low intensities a number of questions or difficulties can arise: (i) the X-ray equipment at TWI has a maximum output of 160kV and is not specifically designed for low kV X-rays, but it is being used at the lower extreme of its operating range to output intensities as low as 10kV; (ii) for low kVs the exposure time may be unnecessarily large; (iii) the equipment may indicate a low kV reading on the dial which may be different from the actual output, and which may also vary from set to set; (iv) the output may vary each time the set is switched on; (v) the absorption by the air between the source and specimen may be comparable to the absorption by the test specimens, particularly for thin specimens (e.g. 5mm thick). With all these issues to consider, the overall question arises as to whether low kV is really necessary to inspect polyethylene pipe?

To answer this important question a wide range of experiments were repeated using different thicknesses of polyethylene and at different intensities. These experiments were also used as a check on the earlier exposure charts (11). It was found, for example, that using X-ray energies of between 16kV and 26kV, the exposure times were kept to the order of ½–1 minute. Below 16kV, the exposure times required to achieve the radiograph densities in the range specified by BS 2600 (13), were unnecessarily large. Above 26kV the exposure times are too short to achieve reproducible density readings on the radiographs. So it is clear from these experiments that low kVs are necessary, but perhaps not as low as 10kV.

To check the repeatability of the output intensity of X-ray tubes designed for relatively high X-ray energies, but working at low X-ray energies, radiograph density measurements were taken every 24 hours over a 5 day period on a 20mm thick polyethylene block, with all the other parameters remaining the same. The results achieved were repeatable, with density variations of the order of less than 4%.

The question raised earlier, regarding absorption by air working at low kVs, is considered in Section 4.3. below.

4.2. New Exposure Charts

The range of experiments conducted to assess whether low kV was necessary were also used to check previous exposure charts and to produce updated ones with tighter tolerances on required radiographic densities. The new charts replace those in (11), and now a finer focal spot size is used, in line with a shorter focus-to-film distance. The two new exposure charts illustrated in Fig.28 and Fig.29 cater for the following groups respectively:

Group (a)

Dowlex 2343 E	white
Finathene 3802 B	black
Eltex TV B 121	black
Eltex TU B 124	dark blue

Group (b)

Finathene 3802	light blue
Finathene 3802 Y	yellow
Eltex TU B 125	orange

Further information regarding these polyethylene materials is given in Table 1. For a different polyethylene to those in group (a) and (b), it will be necessary to produce a separate exposure chart in accordance with BS 2600 (13). To do this, step wedges covering the thickness range of interest will need to be manufactured (see (11) for further details). It has not been confirmed why the polyethylene materials considered in this study fall into these two groups, with regard to their absorption of X–rays. This variation in X–ray absorption is in contrast to the way the materials respond to ultrasonic energy. Whilst there are variations in ultrasonic wave velocities and attenuation, these are not sufficiently large to require separate calibration blocks for setting sensitivity (11). There does not appear to be any correlation between the division of the polyethylene materials for radiography and their product specifications given in Table 1. The colour of each polyethylene material is achieved through the addition of certain compounds (e.g. 2% carbon black). These compounds may contribute to the absorption characteristics of each polyethylene material, but further work will need to be carried out to confirm this.

It is worth noting that the above exposure charts were produced from step wedges manufactured from flat plaques, using the same polyethylene pellets as for the pipes. In order to use the above exposure charts for inspecting pipes, it is important to confirm that the plaques do not have significantly different X–ray absorption properties than the pipes. A number of experiments were carried out to confirm this. For a given polyethylene material and thickness (e.g. Finathene 3802 Y), the plaques were 1–2% more absorbent of X–rays than the pipes, that is, the radiographic densities for plaques are 1–2% lower than radiographic densities for the corresponding pipe material.

Absorption of X–rays in metals is determined by the material density and atomic structure, with the effects of material density dominating. Here, the way in which plaques and pipes are manufactured is believed to influence the observed differences in the absorption characteristics. The plaques are manufactured by compacting the polyethylene pellets while the pipes are extruded from the pellets. However, the observed differences in densities are not significant and the exposure charts in Fig.28 and Fig.29 can be used for pipes falling in groups (a) and (b) above.

4.3. Absorption of Low kV X-Rays by Air, Oxygen and Helium

The need to use low intensities, as outlined above in Section 4.1, raises the question whether a medium other than air is more suitable. The absorption of X-rays by air can present problems for intensities less than 20kV. For example, air may have a comparable absorption to the thinner polyethylenes of interest. In order to assess the absorption of air in relation to other media, experiments were conducted above and below 20kV, using helium and oxygen filled balloons with the same focus-to-film distance (see Fig.30). The results are illustrated in Fig.31 and confirm, as expected, that helium absorbs less radiation than air, which in turn absorbs less than oxygen. This does suggest that helium might be a better medium than air when using intensities less than 20kV, as it will have less of an influence over the production of exposure charts. However, further experiments have shown that air absorbs far less than say a 5mm thick polyethylene plaque over the whole range of intensities in the exposure charts (see Table 3). This demonstrated very clearly that the exposure charts are not unduly influenced by the medium, but by the material being inspected. Moreover, repeat inspections of polyethylene pipe, using helium instead of air, have shown that the quality of the radiograph has not improved. Confirming still further that the medium does not significantly affect the radiographic density at low intensities when inspecting polyethylene. It is concluded that, in general, helium is not going to give sufficiently better results than air under these circumstances to warrant using it.

4.4. Image Quality Indicators

Previous work on IQIs considered a comparison between various IQI types manufactured from polyethylene; ASME plaques with holes (ASME V Article 22), step wedges with holes (15), and wire types (12). Of these, the wire types proved to be the most appropriate from the point of view of IQI sensitivity and avoiding the need of numerous spacers normally associated with step/hole type IQIs.

Notwithstanding the difficulties of manufacturing wires from polyethylene with the correct diameter to the specified tolerance, two sets were manufactured (see Fig.32) one set from Eltex TU B124 (intended to be used in conjunction with exposure charts in Fig.28), and one set from Finathene 3802 (intended to be used in conjunction with exposure charts in Fig.29). However, a large number of experiments have shown that either of these wire types can be used with both exposure charts, and that it is not necessary to use two separate sets for the two polyethylene groups (a) and (b) outlined here in Section 4.2.

One outstanding problem with polyethylene wire type IQIs is in their handling. Currently they are not encapsulated with any protective cover and, as Fig.32 and (12) shows, the vast majority of wires are less than a millimetre in diameter and can be damaged easily. To overcome this problem a number of materials were considered as encapsulation materials. These materials must protect the polyethylene wires, but must not significantly absorb X-rays and unduly influence the final radiograph. The following materials have shown the most promise:

Cling film (used for wrapping food);
Poly Pockets (polyethylene pockets used for paper documents);
Polyethylene material for encapsulating metal IQIs.

This appears surprising at first, particularly as it involves using polyethylene material. These materials are thinner than the wires themselves and so do not absorb a significant amount of radiation. This is in contrast to paper, which has also been evaluated, and while it is also thin, its grainy structure is visible at low kVs and has the effect of masking the wires.

With the above materials, both single and double layers have been considered and radiographic densities between 1.8 and 3.0 have been achieved with 6 wires visible in all cases. A double layer of cling film looks the most promising at this stage.

In preparation for a standard for testing polyethylene pipes, a table of IQI sensitivity has been produced (Table 4), covering the polyethylene thicknesses of interest here, and indicating the minimum number of wires that should be visible on the radiograph. The IQIs used to produce Table 4 are those illustrated in Fig.32(c). Table 4 considers normal and critical techniques and represents what is possible under laboratory conditions. In an industrial environment it is anticipated that the number of wires visible may be relaxed. However, the work carried out to produce Table 4 has been repeated a number of times and is considered to be an accurate and good starting point for the discussions on a standard for polyethylene IQIs for testing polyethylene pipes.

4.5. The Detection of Nominally Cold Welds and Their Significance

As mentioned previously in this report, the term 'cold weld' is used to describe a weld between two sections of polyethylene pipe whose heated ends have begun to crystallise before being joined together. Nominally cold welds can occur if, during the butt welding process, either the hot plate temperature is too low or the time between removing the hot plate and joining the pipe ends is too great. Characteristically such welds tend to fail in a brittle manner at a tensile strength lower than that of a sound weld. For the particular grade of polyethylene considered in this study (Finathene 3802Y), work at TWI has shown that, under laboratory conditions, a nominally cold weld is associated with plate removal times of the order of 40 seconds and above (4). However, recommended welding parameters for polyethylene pipe (5) stipulate a maximum hot plate removal time of 4 seconds.

One of the objectives of this study was to see if any useful quantitative relationship exists between the radiographic density and the delay time between removal of the hot plate and joining of the pipe ends. In extreme cases it is envisaged that this information could be used to infer the presence of a nominally cold weld.

In the light of the new exposure charts it was decided to radiograph the four pipe specimens, fabricated using a range of plate removal times, in a more systematic and consistent way (i.e. keep all equipment settings the same and carry out the radiography on the same day if possible). In this way the radiographic density of a standard pipe (i.e. welded under optimum conditions), measured at a number of positions around the pipe's circumference, could be compared with the radiographic densities at identical positions on the pipe specimens welded with various plate removal times. The results are illustrated in Fig.33, where a standard pipe (H59-3) is compared with a number of pipe welds manufactured with delay times of 20 seconds (7075/H2), 30 seconds (7075/H3), 40 seconds (7075/H4) and 50 seconds (H59-2). The increase in radiograph density, corresponding to an increase in plate removal times, is clear for the 40 second and 50 second delay cases. The 20 second and 30 second delay cases are not too dissimilar to the standard pipe.

The mean radiographic density (over the whole circumference) for each of the above pipes is plotted against plate removal time, and is illustrated in Fig.34. Whilst the relationship in Fig.34 is not a simple linear one, there is scope to confirm an empirical relationship between density and plate removal time.

It has always been recognised that the general increase in radiographic density with increase in plate removal time (as illustrated in Fig.34), is consistent with a thinning of the weld beads. With this in mind, profileometry measurements of the weld beads were carried out on all of the pipes with delay times as considered in Fig.34. The results are illustrated in Fig.35 where the external bead cross sectional area (i.e. height of bead x width of bead) is plotted against the pipe position around the circumference (this 'position' is defined by ranges of 130mm long). With the exception of the 20 second delay results, the rest show, in general, a clear progressive trend to a decrease of the outer weld bead area with an increase in the delay (plate removal time). The mean bead area (over the whole circumference) for each of the pipes is plotted against the plate removal time and is illustrated in Fig.36, and as might be expected exhibits an almost inverse relationship to that in Fig.34.

Both Fig.34 and Fig.36 suggest that there is scope to infer the time delay between joining of the pipe ends from either mean radiographic density measurements or mean profileometry measurements of the outer bead. A plot of mean radiographic density and mean outer bead cross-sectional area is given in Fig.37, and suggests a linear relationship may exist between these two parameters. The results in Fig.37 have been derived from Fig.34 and Fig.36 for each time delay.

Another interesting phenomena, discovered with welds made with increased delay times, is that there appear to be some indications or flaws, albeit small, which are not always initially visible on the radiographs. This is particularly true when both beads are left on, and has been demonstrated by removing the beads in a systematic way on the welds 7075/H3 (30 second delay), 7075/H4 (40 second delay) and H59-2 (50 second delay). To illustrate this with the 30 second delay weld, consider Fig.38-40. In Fig.38(a) and (b) a total of two indications are observed on the radiographs. On removing the outer bead and carrying out further X-ray inspection, the two indications previously observed in Fig.38 are still visible (see Fig.39(a) and (b)). However, in Fig.39(c) a 'new' and third indication is observed. On removing the inner bead and carrying out further X-ray inspection, it was found that indications 1 and 3 disappeared, confirming they must have been in the inner bead. However, indication 2 has remained, confirming it is in the weld, and a fourth indication has appeared, which means it too is in the weld (see Fig.40(a) and (b)).

It is difficult to say precisely what these flaws in the beads are, and how they are formed without further analysis.

Whilst the flaw detection performance of X-ray NDT has been demonstrated for polyethylene pipe, what is not clear is the significance of the flaws detected and whether the weld is accepted or rejected. The significance of flaws has been, and is being, considered by other departments within TWI (5,16,17). However, in the context of NDT it was decided to carry out a tensile test on a section of the 30 second delay weld which also had the flaws as illustrated in Fig.40. A tensile test specimen was prepared, according to recommended procedures (16), and the results of the test are illustrated in Fig.41.

At recommended conditions the tensile strength of a good weld for this polyethylene should be $20Nmm^{-2}$, with an associated elongation of 40mm. The performance of 7075/H3, shown in Fig.41, had an elongation of 47mm at a failure force of 13.50kN. The tensile strength for this weld was $23Nmm^{-2}$, suggesting the 30 second delay weld satisfies the recommended conditions of a good weld, notwithstanding the presence of the small flaws. The ductile fracture surface of this weld also indicates a good weld. Whilst this is a consistent result with earlier work (5), further work to establish the precise level of allowable flaws and the degree of nominal cold weld needs to be addressed so that pipes can be inspected to specific accept/reject criteria.

4.6. Updated Procedures for X-Ray Radiography of Polyethylene Pipes

The prototype procedures previously given in (11) have been updated in the light of new exposure charts and the IQI sensitivity table. They are included here in Appendix B for completeness. It is intended to utilise the latest work and submit a proposed standard on the X-ray radiography of polyethylene pipes to the relevant BSI committee for consideration. A document is currently being developed and based on BS 2910:1986 (18).

5. DISCUSSION

The ultrasonic and radiographic techniques considered in this report clearly demonstrate an ability to detect a range of flaw types in butt welded polyethylene pipe systems. However, despite an ability to detect flaws, the relationship between flaw size and pipe lifetime remains undetermined and, at this stage, it is not known whether the flaws detectable by these NDT techniques are of a critical size or not. In order to realise the full potential of polyethylene pipe, it is important that a critical flaw size is determined and that acceptance criteria, to which butt welds in polyethylene can be inspected, are specified. The effect of contamination of the fusion face of butt welds in polyethylene has been investigated by a number of researchers (16,17,19), but this and other work has yet to establish a maximum allowable flaw size for polyethylene pipe operating in a range of common working conditions.

The detection capabilities of the two NDT methods evaluated in this study are displayed in Table 5. Lack of fusion type flaws, simulated for the purposes of this study by aluminium foil discs introduced at the weld interface prior to fusion, are detected by both ultrasonic and radiographic techniques. Ultrasonic tandem and TOFD techniques have shown that they are able to detect reliably a lack of fusion down to 2mm in through-thickness height. Low kV radiography has demonstrated a similar level of sensitivity to these flaws but, in addition, is sensitive to gross particle contamination of the fusion face (chalk dust).

Ultrasonic tandem and TOFD images of pipe H59-5 display a variation in signal between areas known to contain chalk dust and areas that are dust free. However, further work is needed to establish whether this variation in signal is truly due to the chalk dust or whether it is due to another feature of the weld, such as a variation in bead condition for example.

Perhaps the most significant benefit in detection performance when using radiography is its ability to detect a weld made with a long plate removal time and, in the most extreme circumstances, the presence of a nominally cold weld. To check for this is relatively straightforward. The density of a radiograph, produced during a routine volumetric inspection of a butt weld in polyethylene, can be measured at positions along the weldline and compared with Fig.34 to determine the interval between removal of the hot plate and fusion of the pipe ends. In this manner it is envisaged that the presence of a nominally cold weld can be detected.

The ultrasonic techniques considered in this study did not appear to be sensitive to nominally cold welds. However, the practical application of an ultrasonic technique is arguably more straightforward than radiography, and should not be overlooked as a means of inspection. Unlike radiography, ultrasonic techniques do not require any special safety procedures. The tandem technique can be used to provide a relatively rapid scan of the weld for volumetric flaws and a separate creeping wave scan can be used to inspect for discontinuities beneath the weld bead. Any irregularities detected during the tandem scan can be marked on the pipe and re-examined using a TOFD technique. TOFD gives improved defect resolution over the tandem technique and can be used to gain an appreciation of the through-thickness height of the discontinuity.

6. CONCLUSIONS

- Ultrasonic creeping waves can be used to inspect the fusion area just beneath the outer weld bead in polyethylene pipes.

- Ultrasonic tandem and time-of-flight diffraction (TOFD) techniques can be used to detect, image and size perpendicular planar flaws in butt fusion joints in polyethylene pipes.

- Ultrasonic techniques are currently unable to detect nominally cold welds.

- Radiographic intensities of between 16kV and 26kV are optimum when inspecting thicknesses of polyethylene pipe between 5mm and 50mm.

- With radiography it is possible to infer and quantify the degree of nominal cold weld and bead size, from local radiographic density measurements.

- Polyethylene wire type image quality indicators (IQIs) have been proven to achieve radiographic sensitivities at least as good as polyethylene plaques and step wedges with holes, and have other practical handling advantages with regard to pipe inspection.

7. ACKNOWLEDGEMENT

The work was carried out within the Core Research Programme of TWI which is jointly funded by the Information and Manufacturing Technology Division of the Department of Trade and Industry, and the Industrial Members of TWI.

The authors are grateful to Kim Hayward, Jeff Dobbs and Denis Fish for their help in preparing this report.

8. REFERENCES

1 Gwartney Jr. W R: 'Detecting flaws in PE pipe butt fusion joints'. T D Williamson Inc. trade literature. February 1987.

2 Dodin M G: 'Welding mechanisms of plastics - A review'. J. Adhesion, 12 (1981).

3 WIS No.4-32-08, August 1988: 'Specification for site fusion joining of MDPE pipe and fittings'.

4 Watson M N: 'The butt fusion (hot plate) welding process for polyethylene and polyethylene pipe - Improvements to productivity'. TWI Members report no. 419/1990.

5 Watson M N: 'A study of the butt fusion (hot plate) welding process for polyethylene pipe'. TWI Members report no. 382/1988.

6 Ewing L and Richard W: 'Polyethylene gas pipe systems: An appraisal of joint design and construction methods'. British Gas report no. ERS E.Z67, November 1980.

7 Troughton M J and Booth G S: 'Structural integrity of butt fusion welded polyethylene pipes - a review'. Draft TWI Members report, March 1994.

8 Gas Research Institute: 'Quality assurance of heat fused polyethylene plastic pipe joints'. GRI 85/0284. Final report November 1983-June 1986.

9 Werner Roye: 'X-ray testing of thermoplastics welding joints'. Proc. 4th European conference on NDT, London 1987, Vol.3, 2051-2055.

10 Edwards G R: 'The development of non-destructive test procedures for hot plate butt fusion welds in polyethylene', TWI Members report no. 370/1988.

11 Georgiou G A and Macdonald I A: 'Ultrasonic and radiographic NDT of butt fusion joints in polyethylene'. TWI Members report no. 465/1993.

12 BS EN 462-1:1994: 'Non-destructive testing - Image quality of radiographs'. Part 1: Image quality indicators (wire type) - Determination of image quality value.

13 BS 2600:Part 1:1983: 'Radiographic examination of fusion welded butt joints in steel'. Methods for steel 2mm up to and including 50mm thick.

14 Langenberg K J, Fellinger P and Marklein R: 'On the nature of the so-called subsurface longitudinal wave and/or the surface longitudinal "creeping wave"'. Res. NDE (1990) 2, 59-81.

15 BS EN 462-2:1994: 'Non-destructive testing - Images quality of radiographs'. Part 2: Image quality indicators (step/hole type) - Determination of image quality value.

16 Sims J: 'The butt fusion (hot plate) welding process for polyethylene pipe of 125 and 400mm diameter'. TWI Members report no. 434/1991.

17 Giradi M A: 'The evaluation of mechanical testing techniques and the influence of defects in MDPE butt fusion welds', 8th Int. Conf. on Plastic Pipes, Konlingshof, The Netherlands, 21-24 September 1992.

18 BS 2910:1986: 'Radiographic examination of fusion welded circumferential butt joints in steel pipes'.

19 Marshall G P: 'The influence of welding parameters on the toughness of butt fusion welds in MDPE', Advances in Joining Plastics and Composites, Bradford, 10-12 June 1991.

Table 1 Product specification of medium to high density polyethylene materials used in the study

** Product	Colour	Density (g/cm^3)	Yield strength (N/mm^2)	Modulus (N/mm^2)	Hardness Shore D	Application
Dowlex	White	0.937	17	710	60	Heating, installation, irrigation pipes
Finathene 3802 B	Black	0.948	19	700	59	Anticorrosive coatings, cable jackets
Eltex TU B 121	Black	0.958	25	1200	59	Gas pipes, industrial fluid transportation
Eltex TU B 124	Dark blue	0.951	25	1200	60	Water transportation (drinking)
* Finathene 3802	Light blue					Water transportation (drinking)
Finathene 3802 Y	Yellow	0.941	19	700	59	Gas distribution systems
Eltex TU B 125	Orange	0.951	25	1200	59	Gas transportation

* Details not supplied by manufacturer
** These are the Trade Names of the products as supplied by Stewarts and Lloyds Plastics

Table 2 Polyethylene pipe; standard and with manufactured flaws

Pipe reference	Weld condition
H59–1	Standard
H59–2	'Cold' (~50 second delay)
H59–3	Standard
H59–4	Contains different diameter disc inserts (2, 4, 6, 8 and 10mm)
H59–5	Chalk dust added (~ after 8 second delay)
Pipe outer diameter	250mm
Pipe wall thickness	25mm
Pipe material	Finathene 3802 Y (yellow)

Table 3 A comparison of the radiographic densities with and without the presence of a 5mm thick Finathene 3802 Y, using air as a medium

Milli Amps	Time, seconds	Intensity, kV	Radiographic density	
			With 5mm 3802 Y plaque	Without 5mm 3802 Y plaque
1	60	14	1.17	2.17
1	39	16	1.11	2.19
1	30	18	1.38	2.57
1	20	20	1.23	2.17
1	15	22	1.16	1.95
1	13	24	1.23	2.02

Table 4 IQI sensitivity for critical and normal techniques

Specimen thickness, mm	Section A Critical techniques	Section B Normal techniques
	Minimum no. of wires visible	Minimum no. of wires visible
5	7	6
10	5	4
15	5	4
20	5	4
25	5	4
30	5	4
35	5	4
40	4	3
45	4	3
50	4	3

Table 5 The capabilities of ultrasonic and radiographic NDT techniques for detecting defects in butt welded polyethylene pipes

√ - Detectable

(√) - Technique currently being developed and showing some progress

× - Not detectable

* There may be lower limits to the amount of dust contamination detectable using radiography

▲ Near surface flaws only

		Lack of Fusion	Dust Contamination	Cold Weld
RADIOGRAPHY		√	√*	√
ULTRASONICS	Angled compression waves	√	×	×
	Creeping wave▲	√	×	×
	Tandem	√	(√)	×
	Time of flight diffraction	√	(√)	×

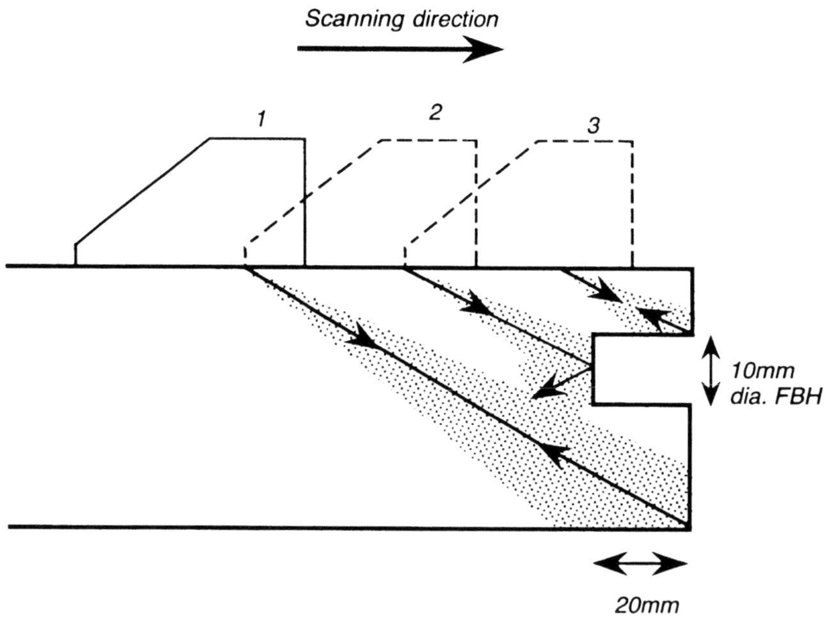

Fig.1 Pulse-echo detection of 10mm diameter flat bottomed hole.

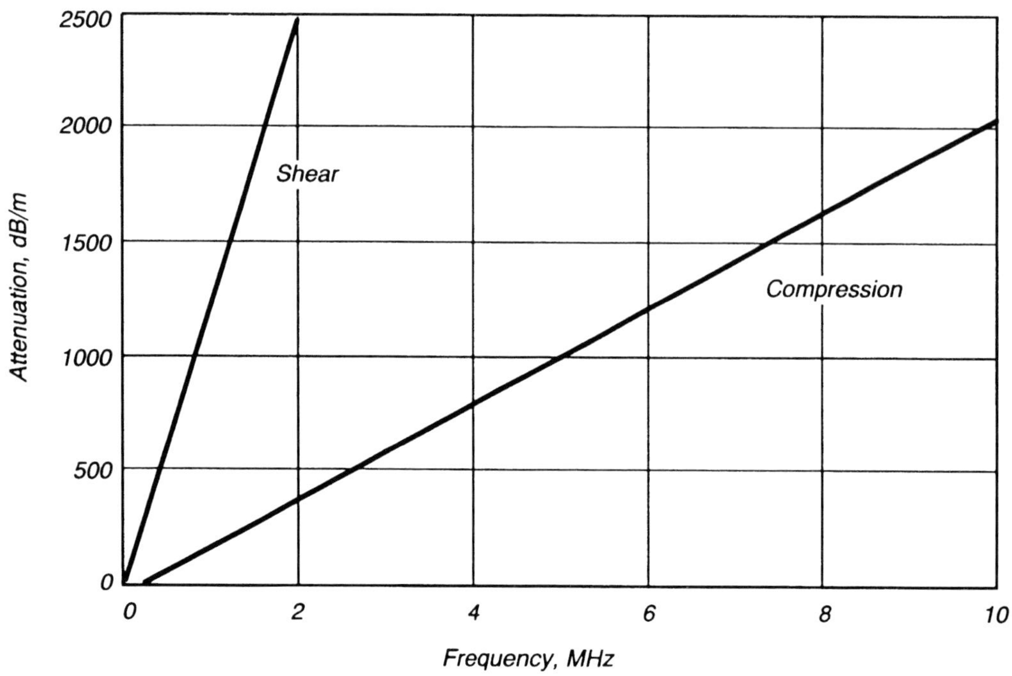

Fig.2 Attenuation of ultrasound of various frequencies propagating through polyethylene at room temperature (from(9)).

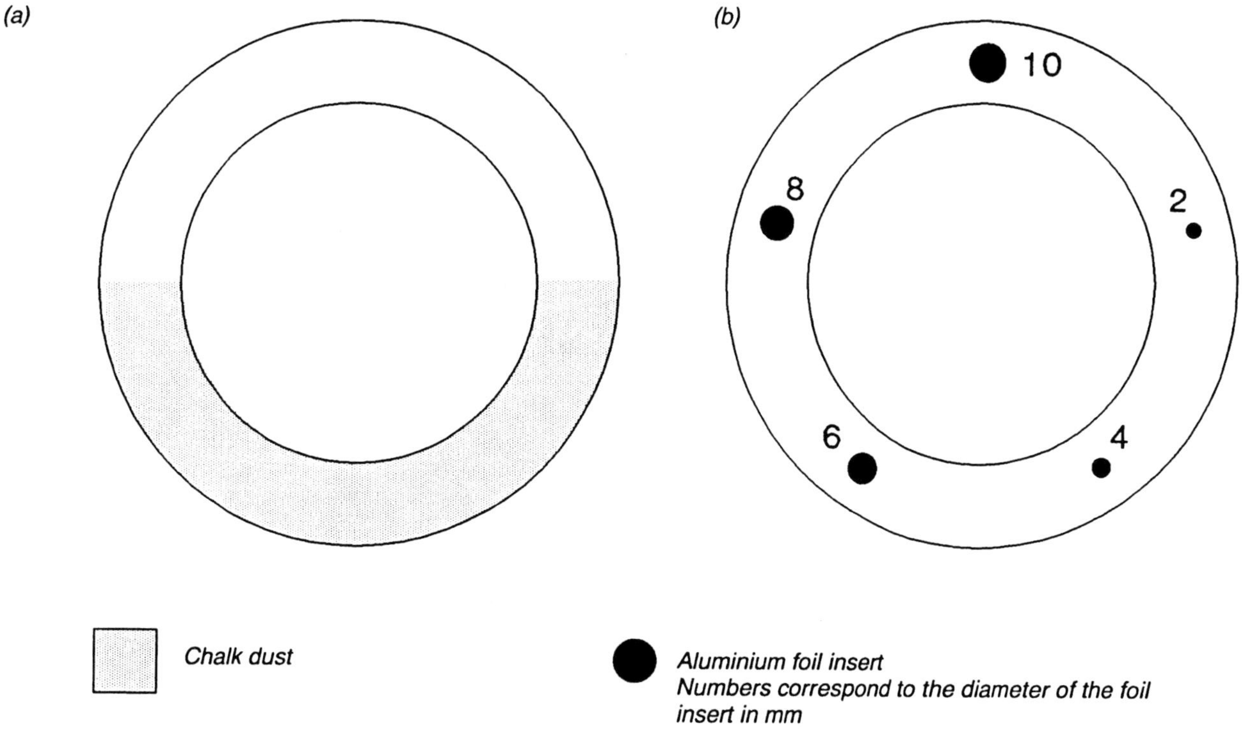

Fig.3 Defect specimens fabricated to include; a) Chalk dust at the fused interface; b) Aluminium foil discs of known diameter at the fused interface.

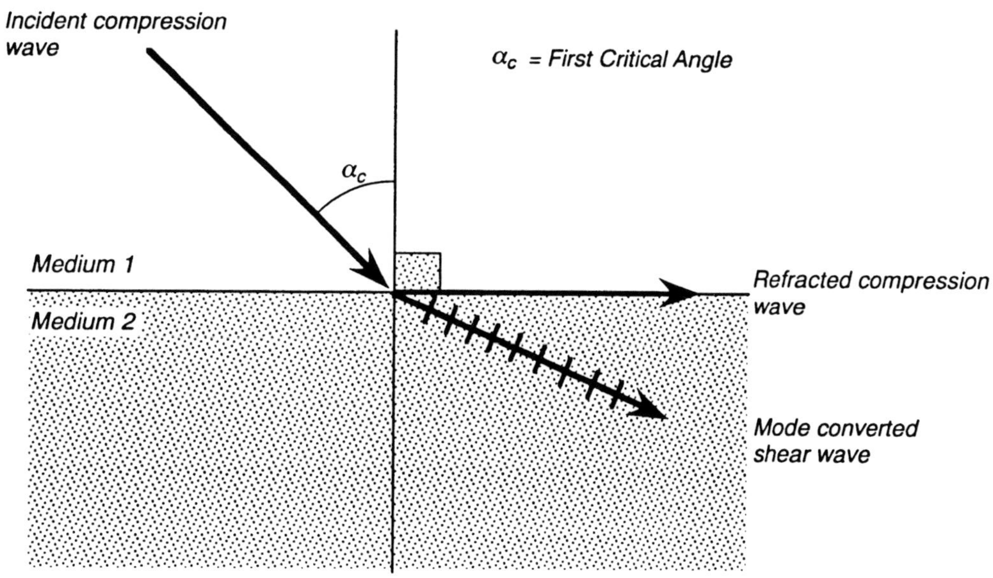

Fig.4 Refraction of compression waves, incident at first critical angle.

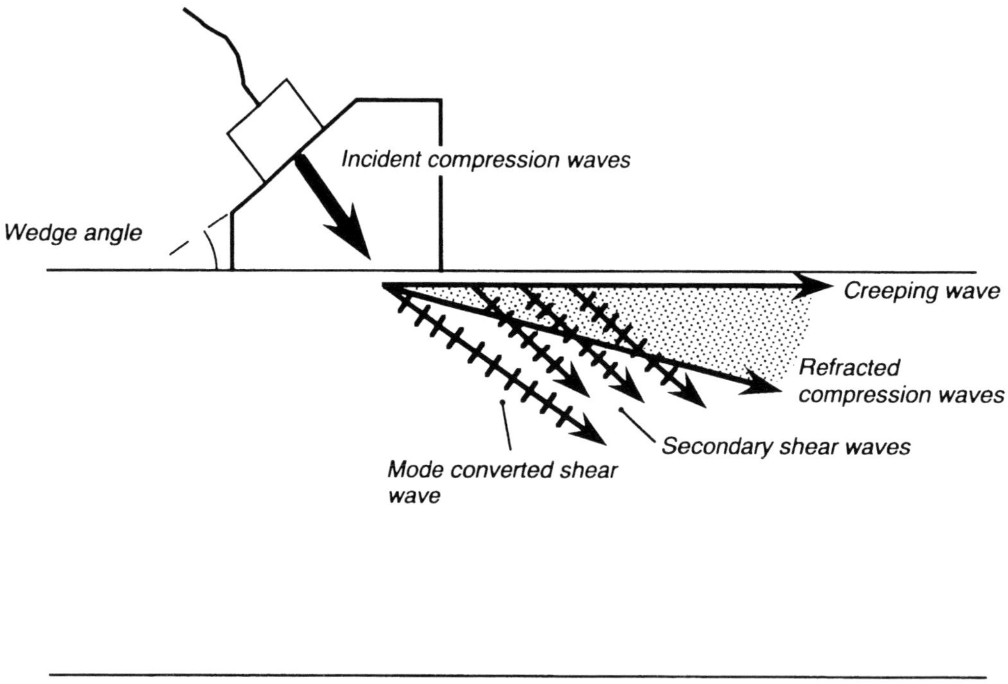

Fig.5 Ultrasonic creeping wave.

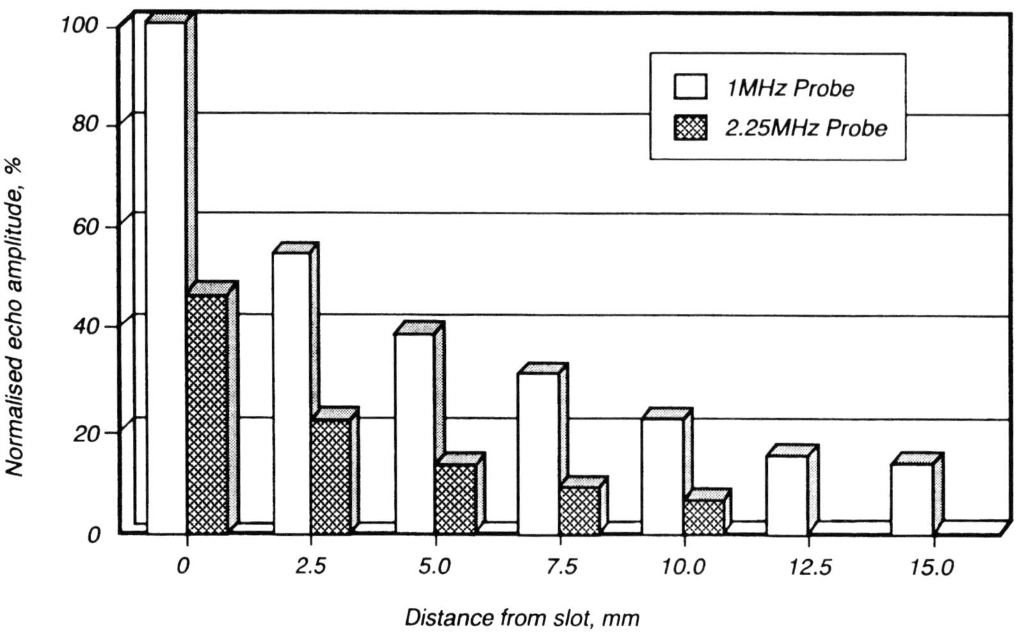

Fig.6 Pulse-echo creeping wave response from 3mm deep saw cut in polyethylene.

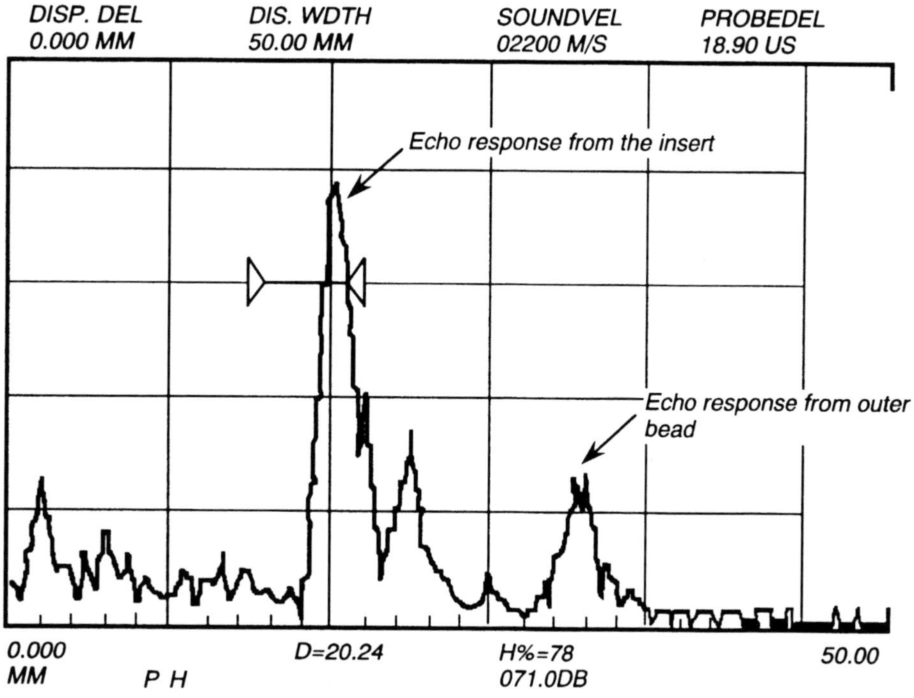

Fig.7 Pulse-echo creeping wave A-scan response from 6mm diameter foil insert.

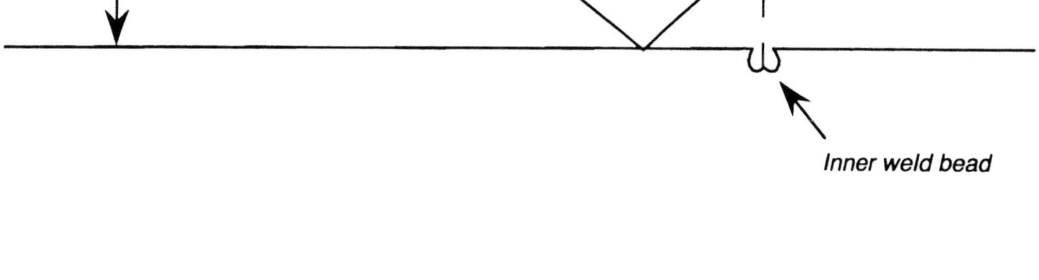

Fig.8 Probe configuration for tandem inspection.

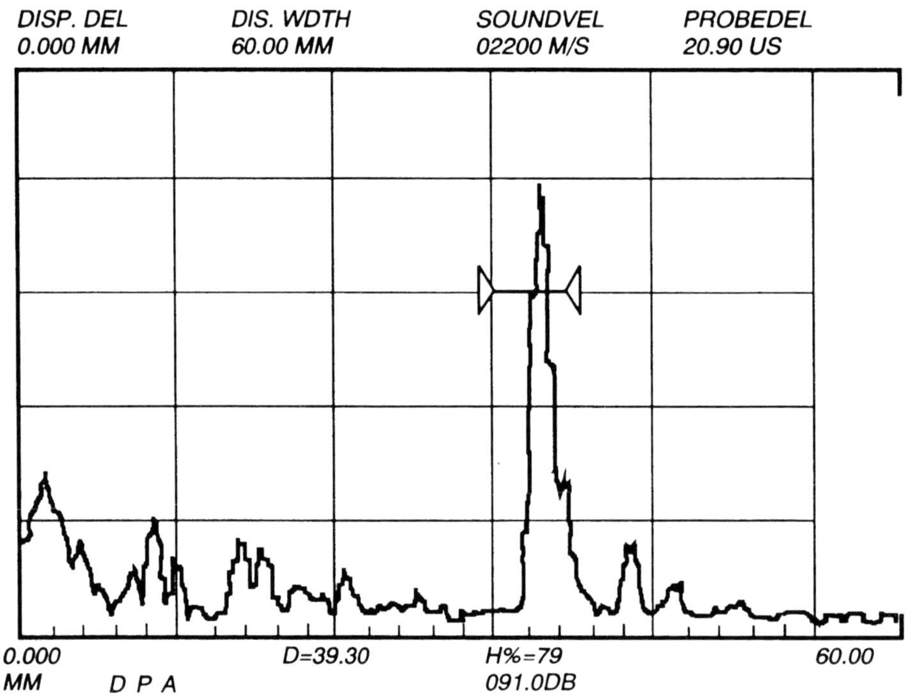

Fig.9 A-scan response frrom perpendicular reflector located in mid-thickness of pipewall, inspected using the tandem technique.

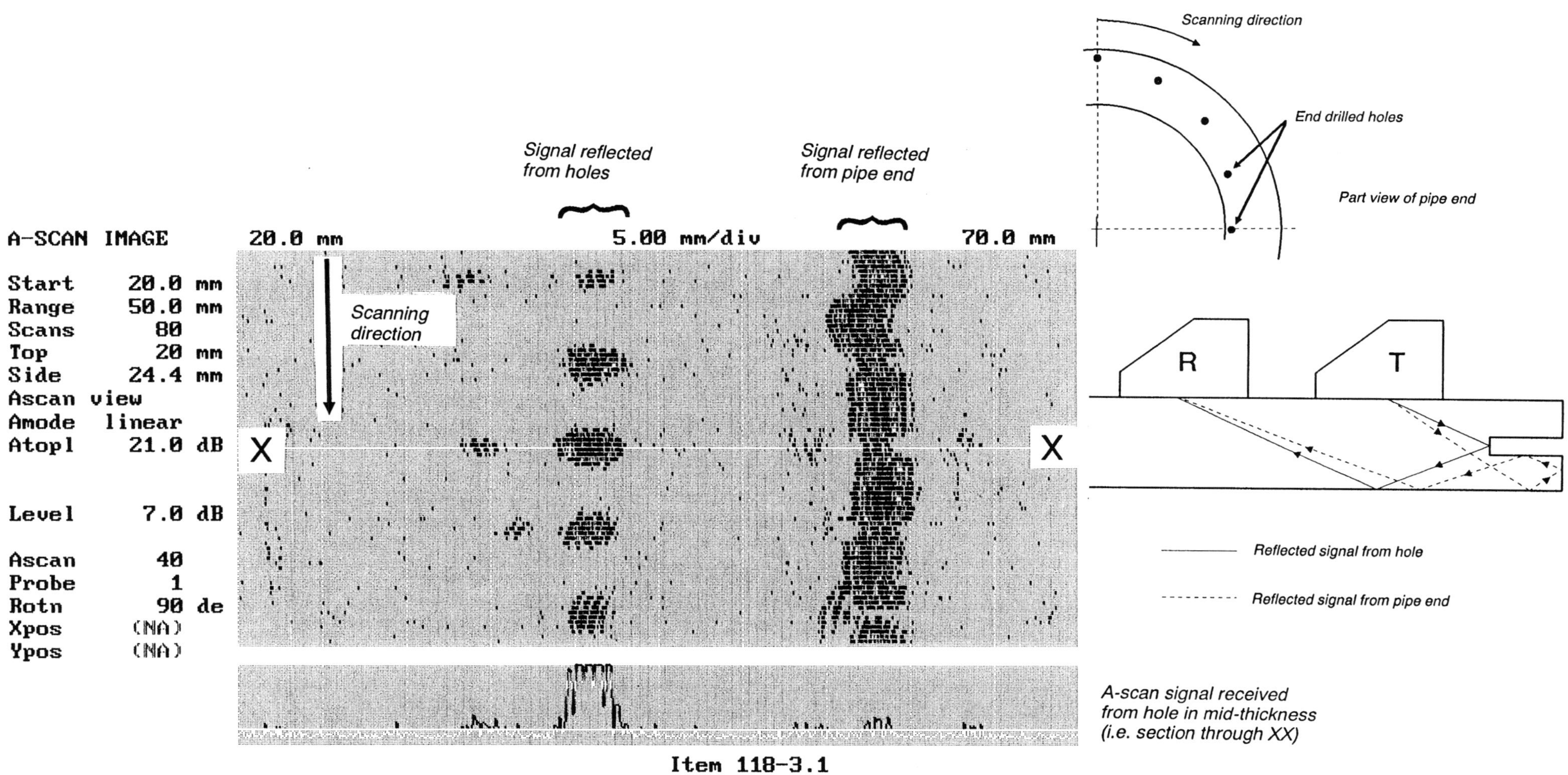

Fig.10 Composite A-scan image of pipe section containing five end drilled holes (scanned using the tandem technique).

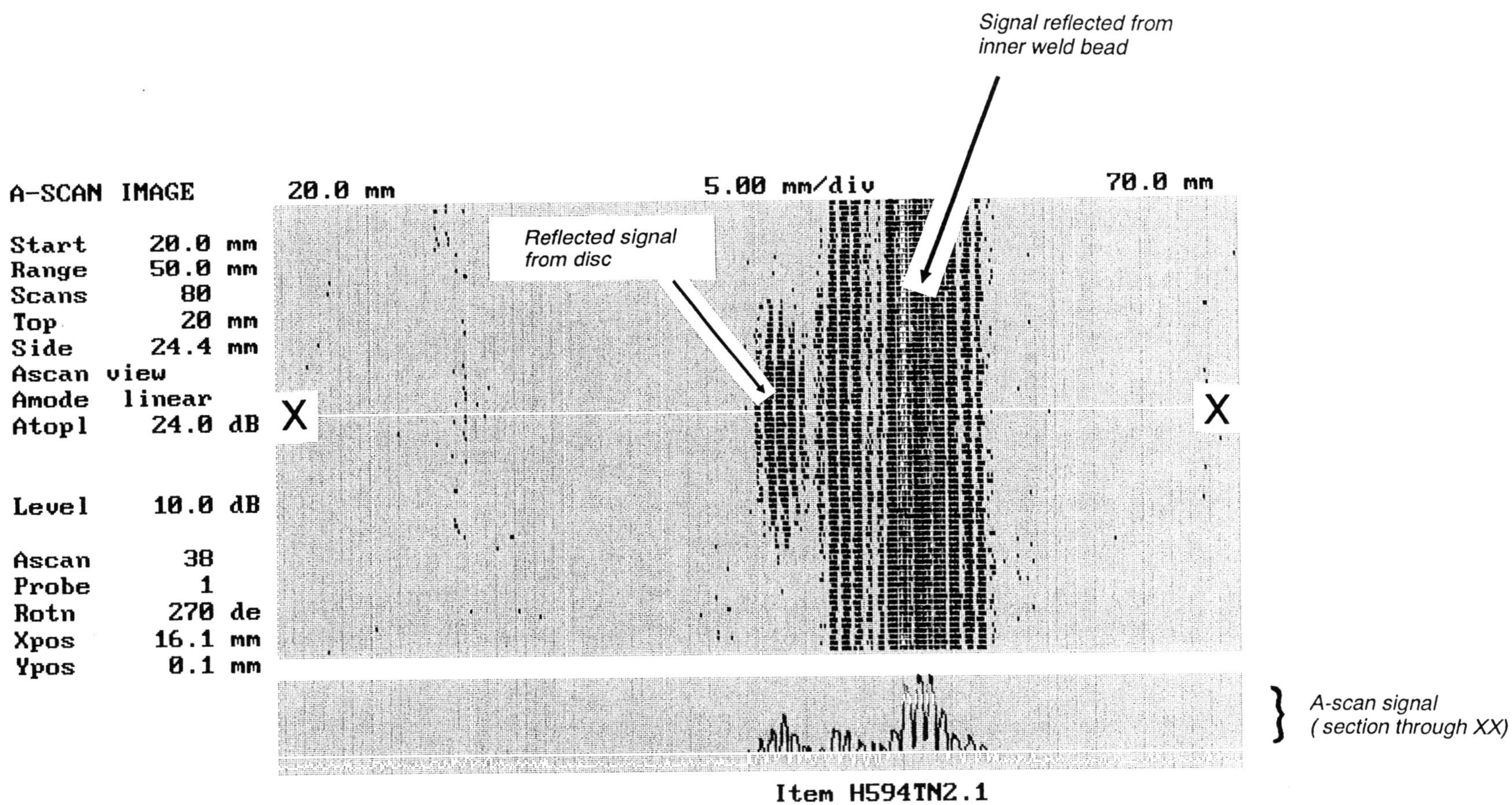

Fig.11 Composite A-scan image of 2mm diameter foil insert (scanned using the tandem technique).

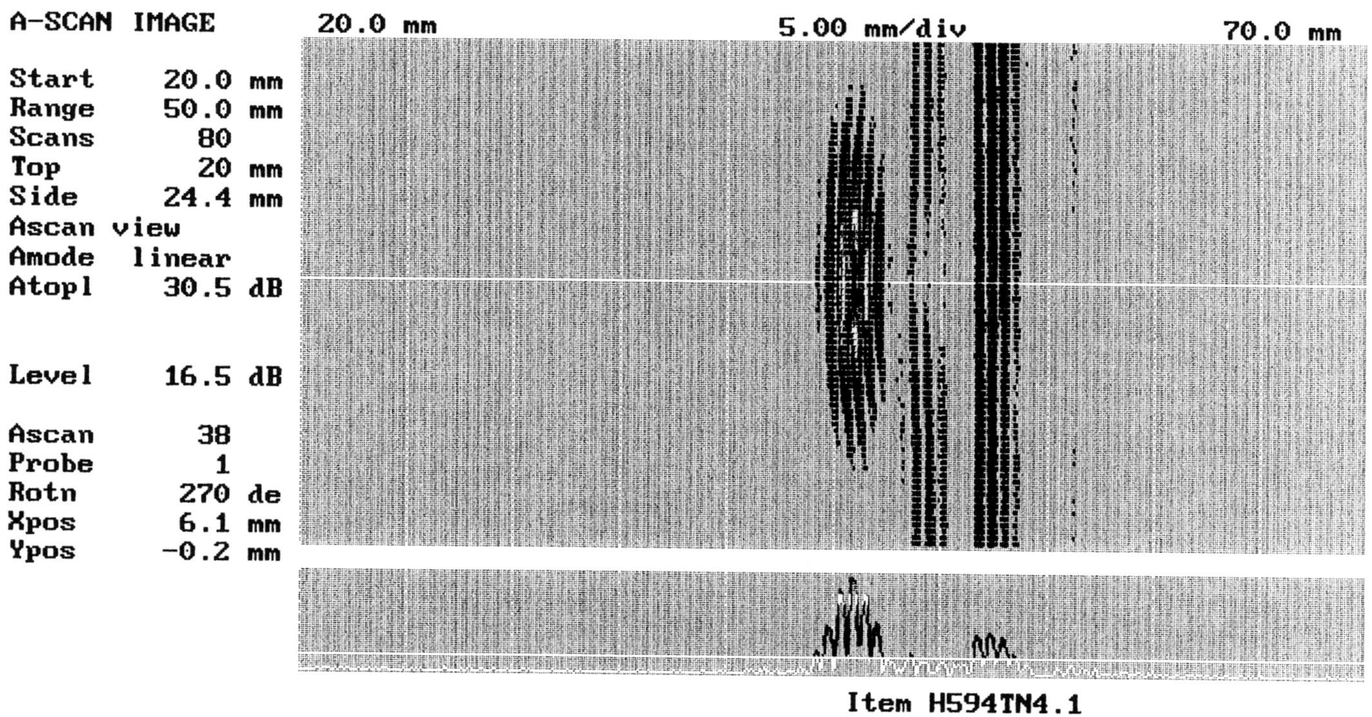

Fig.12 Composite A-scan image of 4mm diameter foil insert (scanned using the tandem technique).

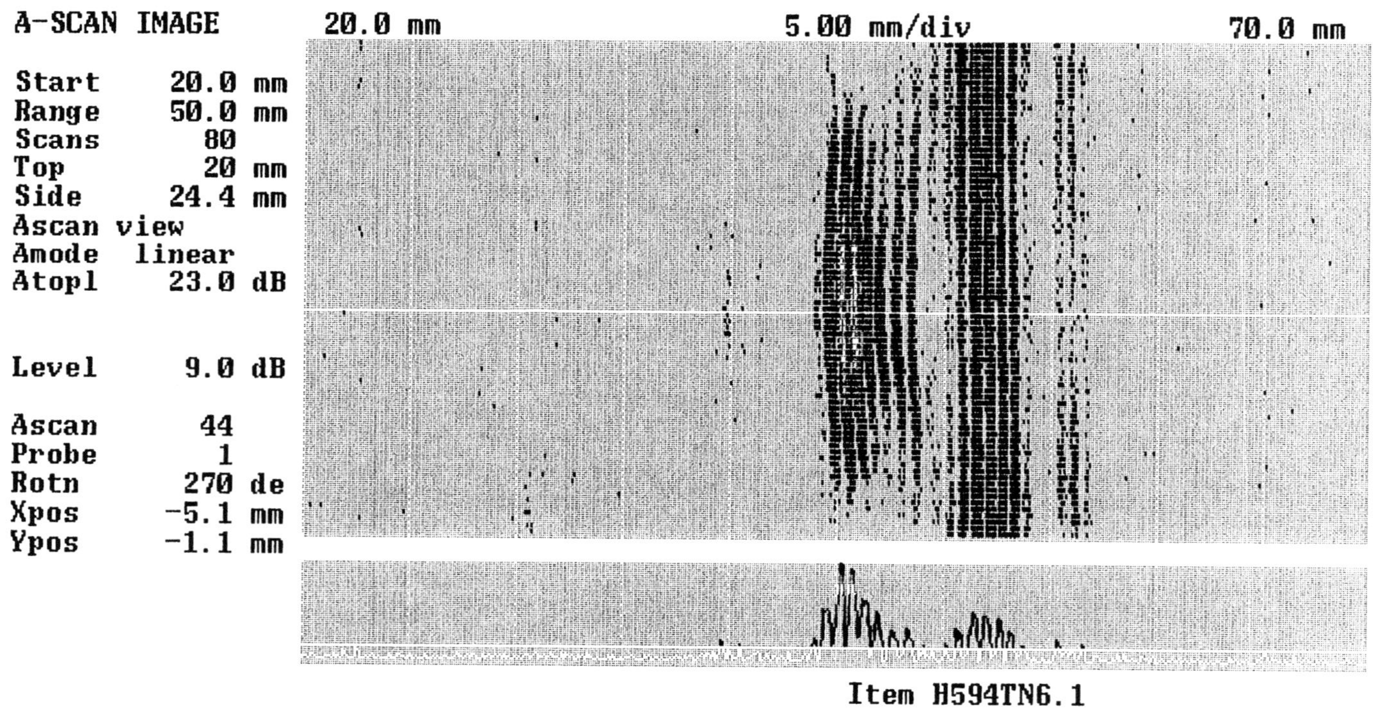

Fig.13 Composite A-scan image of 6mm diameter foil insert (scanned using the tandem technique).

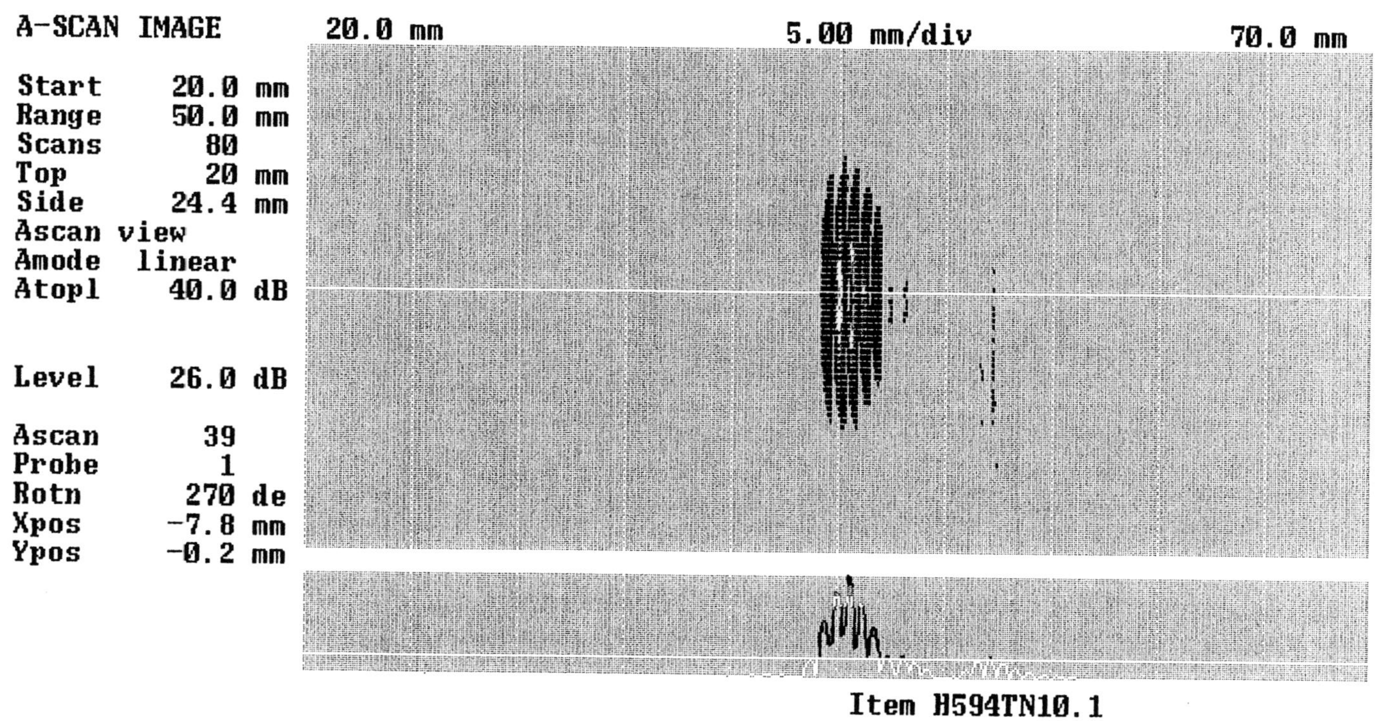

Fig.14 Composite A-scan image of 10mm diameter foil insert (scanned using the tandem technique).

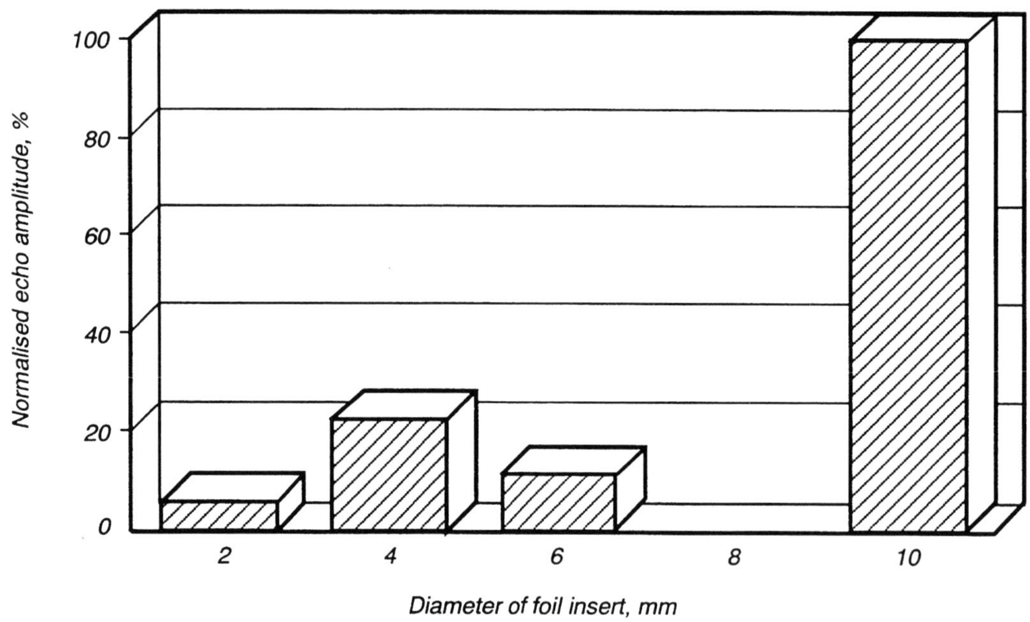

Fig.15 Relative amplitude response from aluminium disc inserts using the tandem technique.

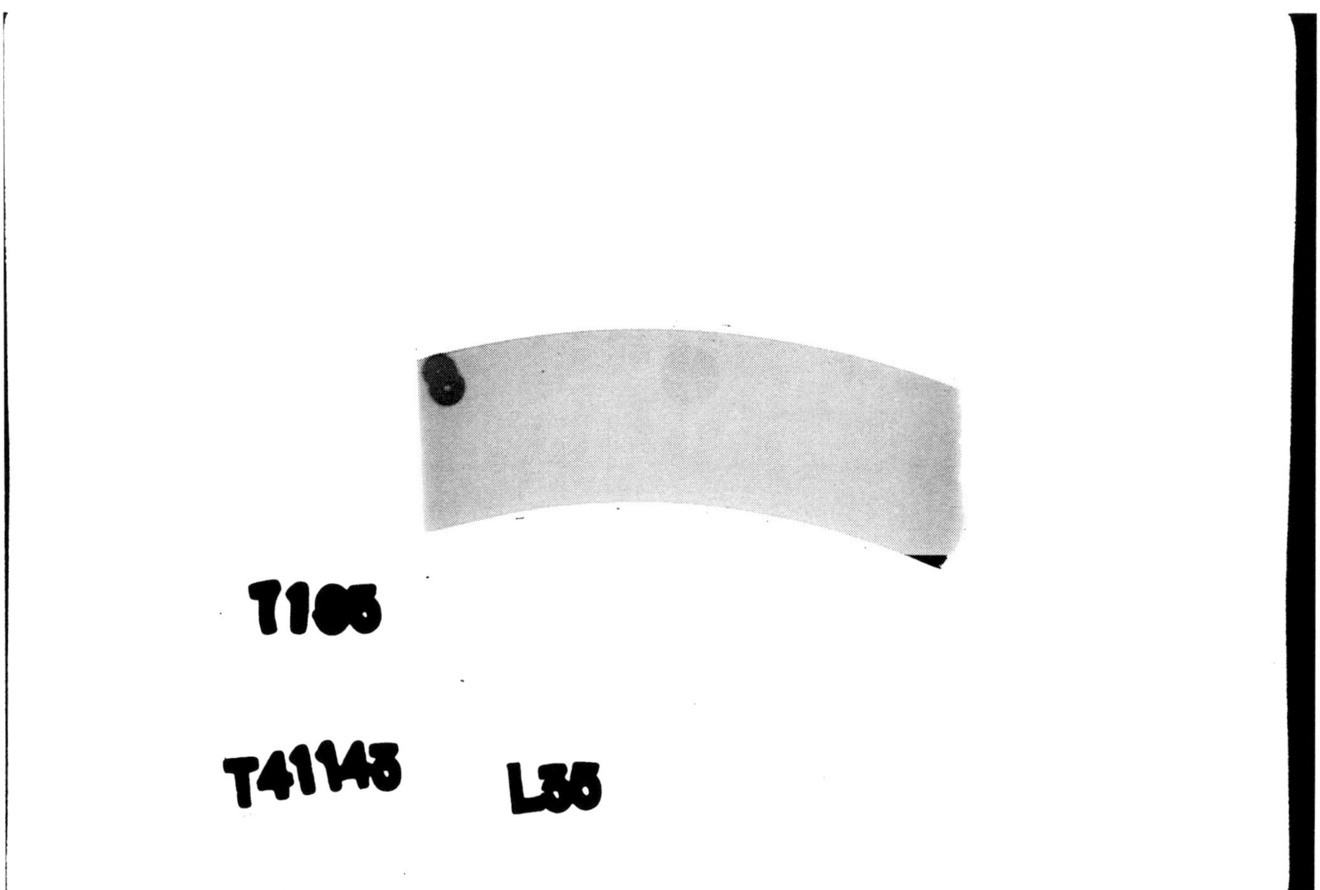

Fig.16 Radiograph showing the position of 8mm diameter foil insert (Radiographic indentification L35).

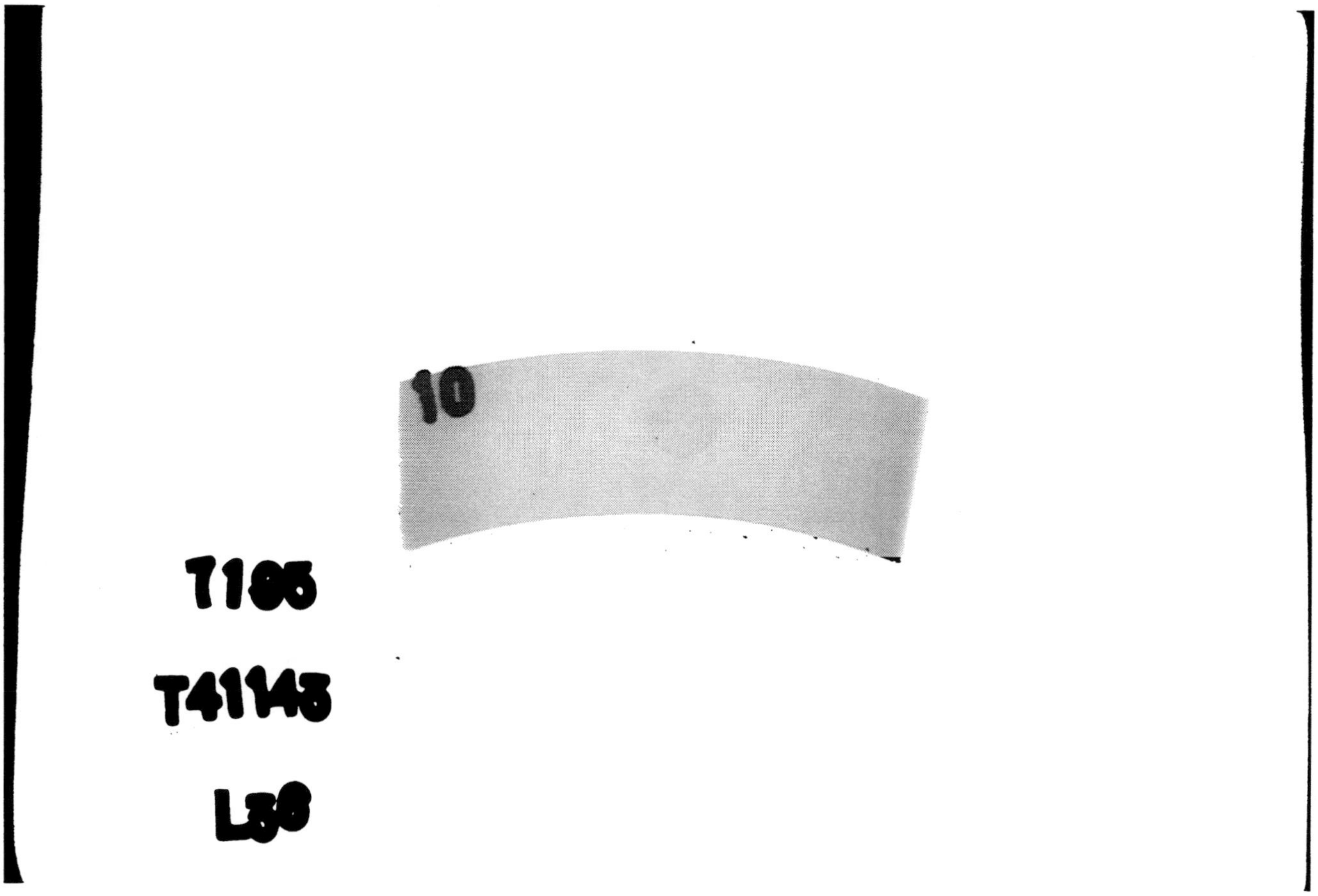

Fig.17 Radiograph showing the position of 10mm diameter foil insert (Radiographic indentification L36).

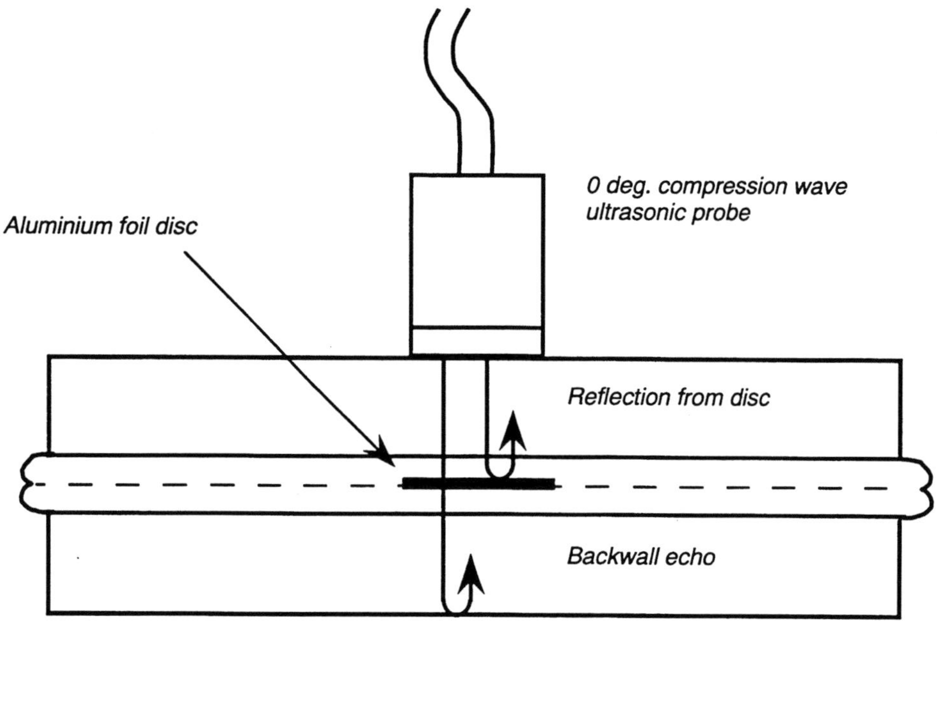

(a)

Fig.18 Normal incidence pulse echo response from sections of pipe containing 10mm and 8mm diameter disc inserts: a) Experimental configuration.

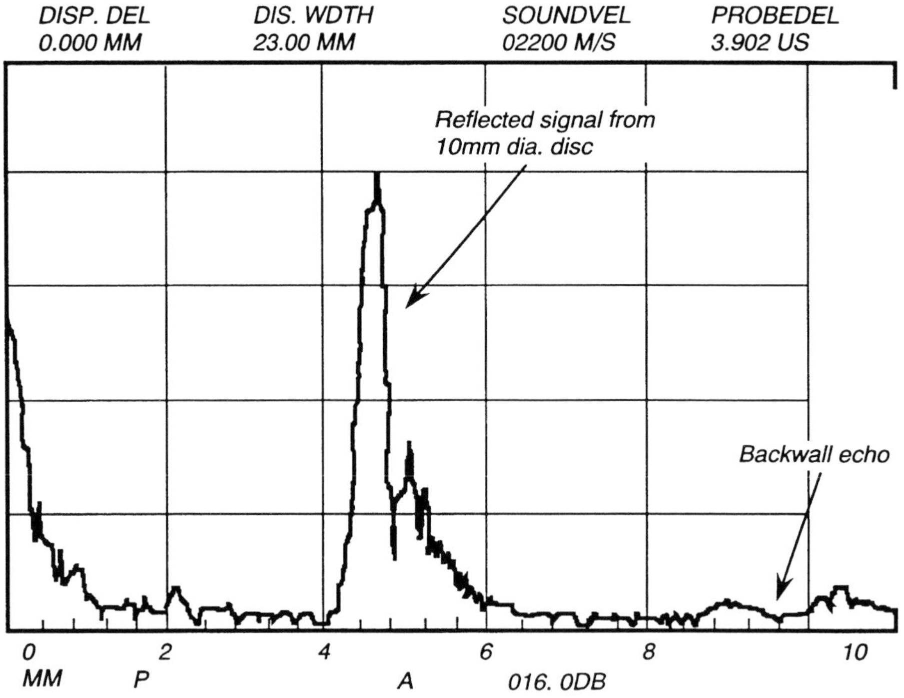

(b)

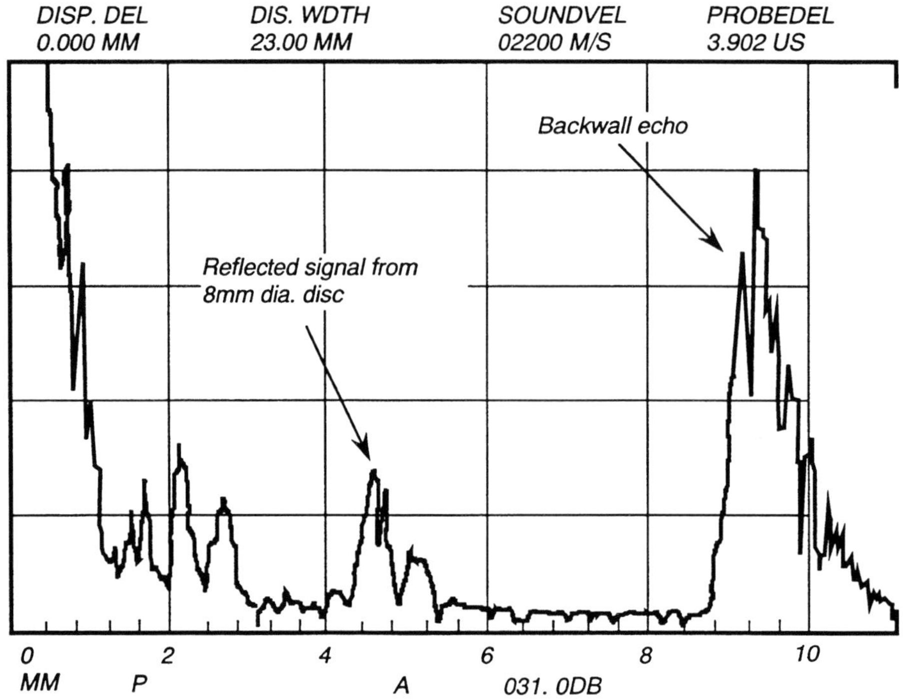

(c)

Fig.18 Normal incidence pulse echo response from sections of pipe containing 10mm and 8mm diameter disc inserts: b) Pulse-echo A-scan response from 10mm dia. disc; c) Pulse-echo A-scan response from 8mm dia. disc.

Acoustic impedence $Z = \rho v$, where ρ is the density of the material and v is the velocity of ultrasound

Reflection coefficient $r = \dfrac{Z_2 - Z_1}{Z_1 + Z_2}$ *(Subscripts 1 and 2 denote the two materials across the interface)*

For 2, 4, 6 and 10mm diameter discs

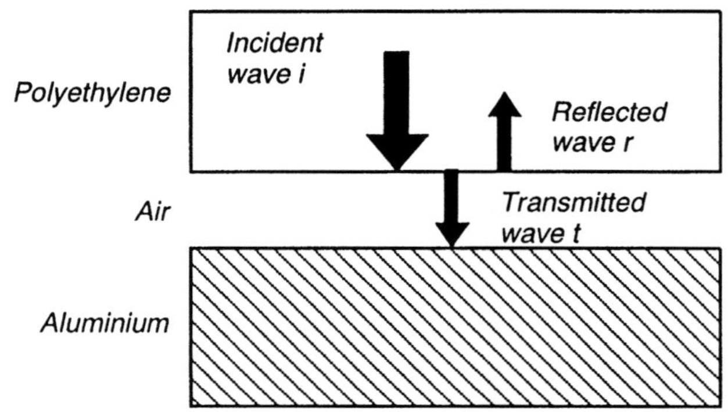

For polyethylene/air interface:

$r = 0.9996$

For the 8mm diameter disc

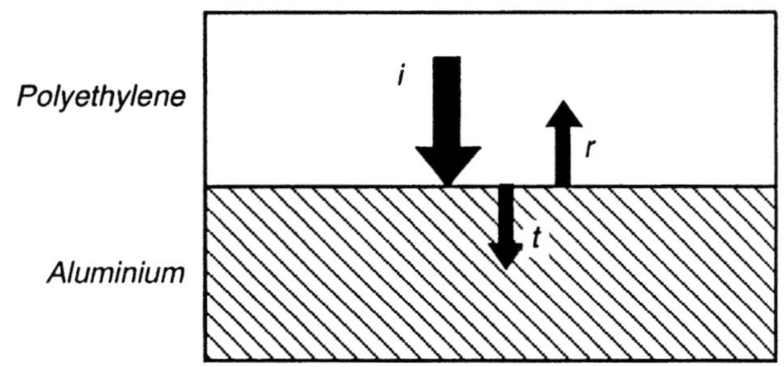

For polyethylene/aluminium interface:

$r = 0.7745$

Fig.19 Assumed interface condition for disc inserts in pipe specimen H59-4.

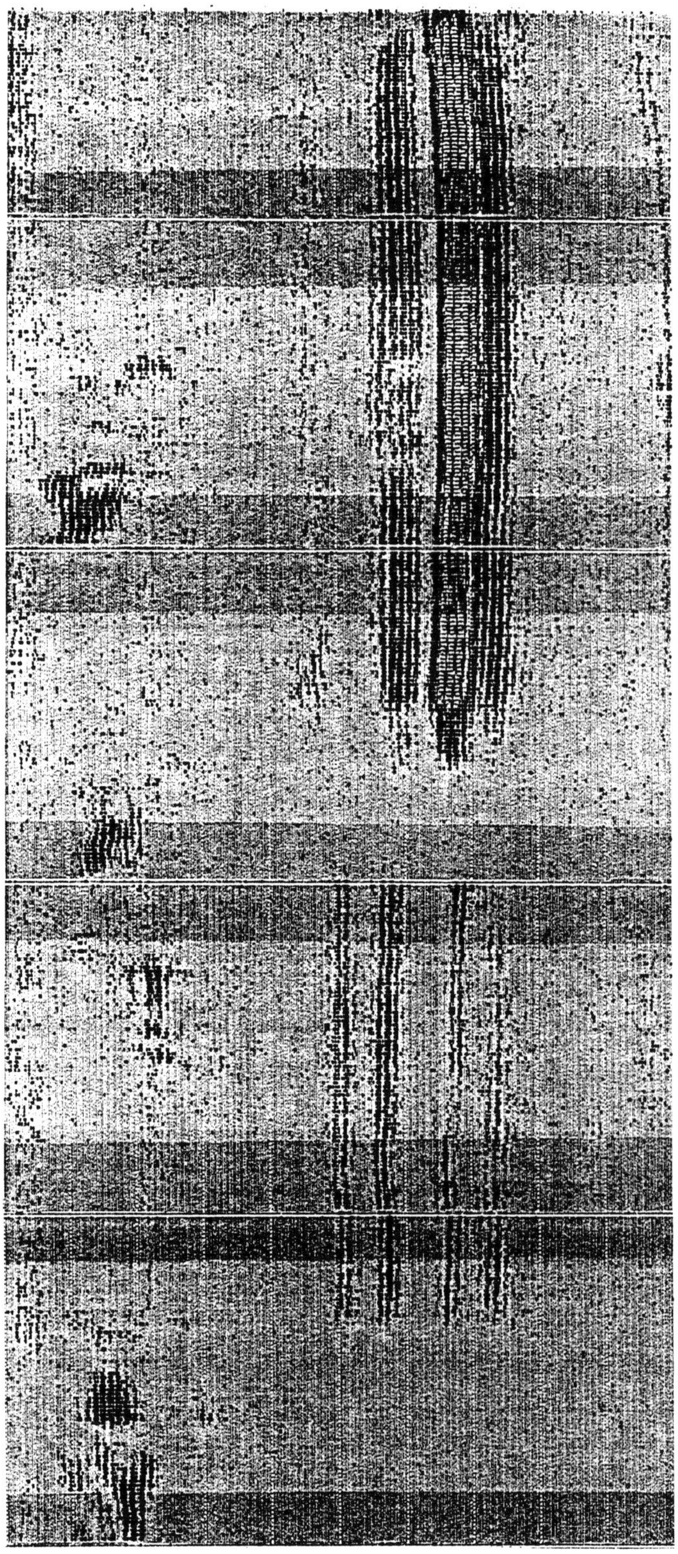

Fig.20 Composite A-scan image of chalk dust (scanned using the tandem technique).

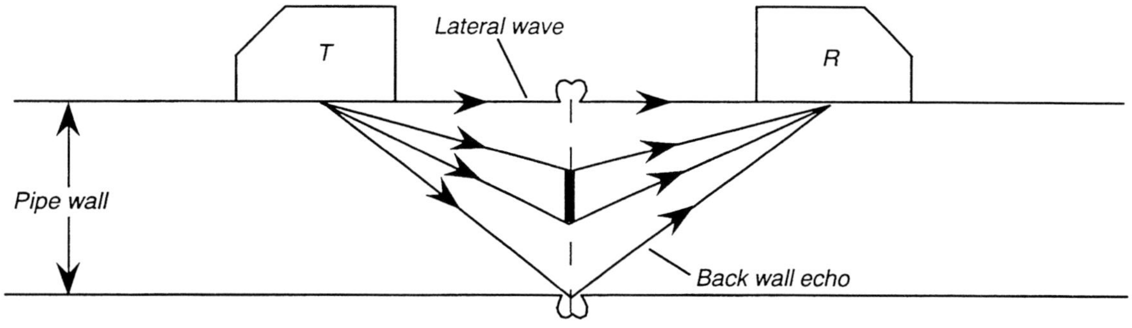

Fig.21 Probe configuration for time-of-flight diffraction (TOFD) inspection.

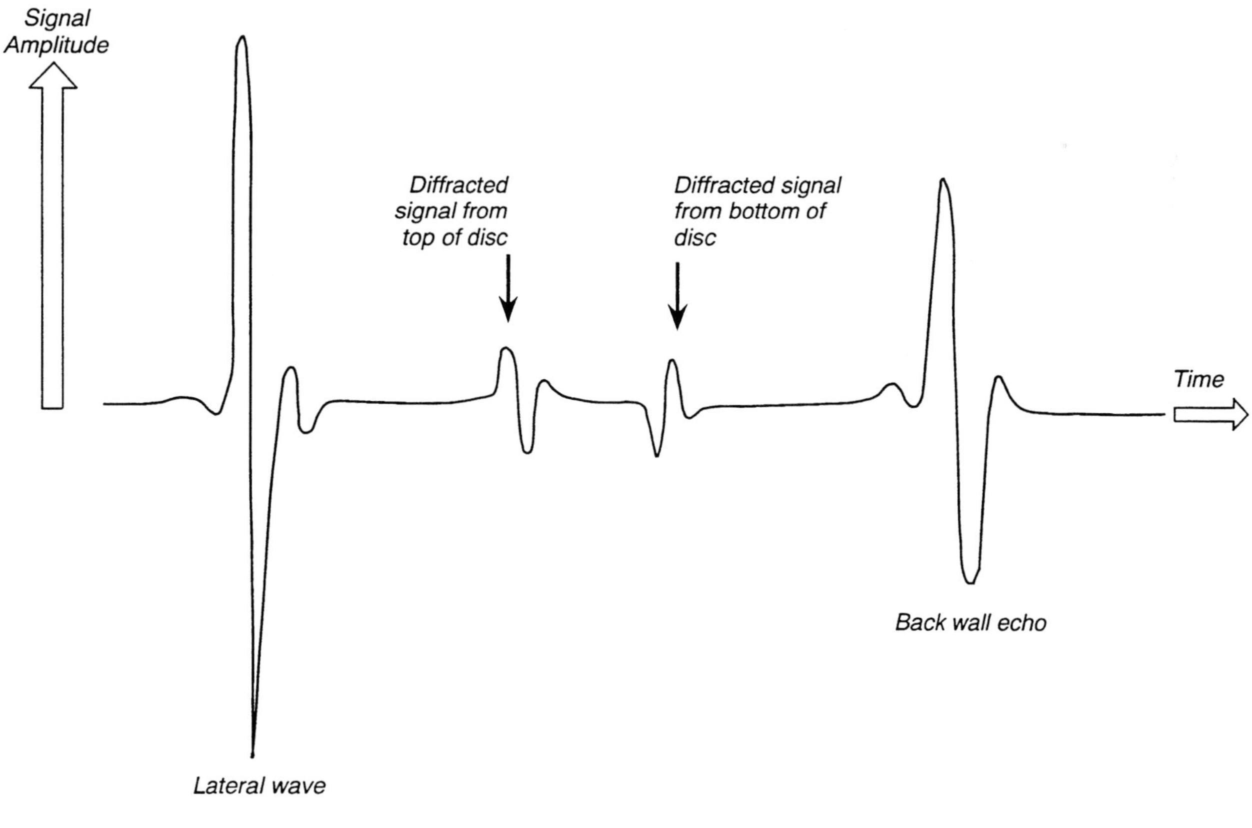

Fig.22 Typical A–scan response from perpendicular defect inspected using TOFD technique.

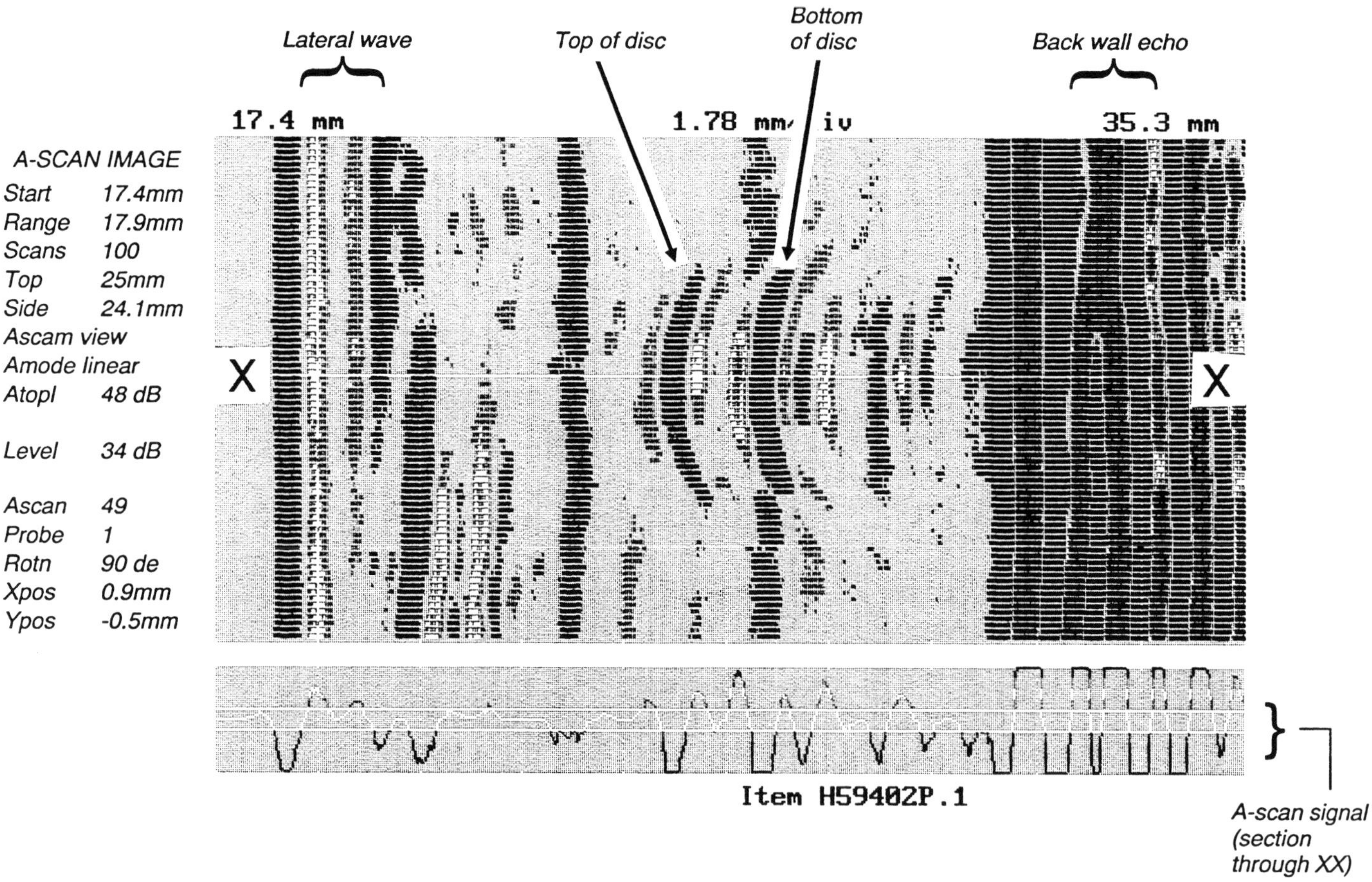

Fig. 23 TOFD image of 2mm diameter foil insert.

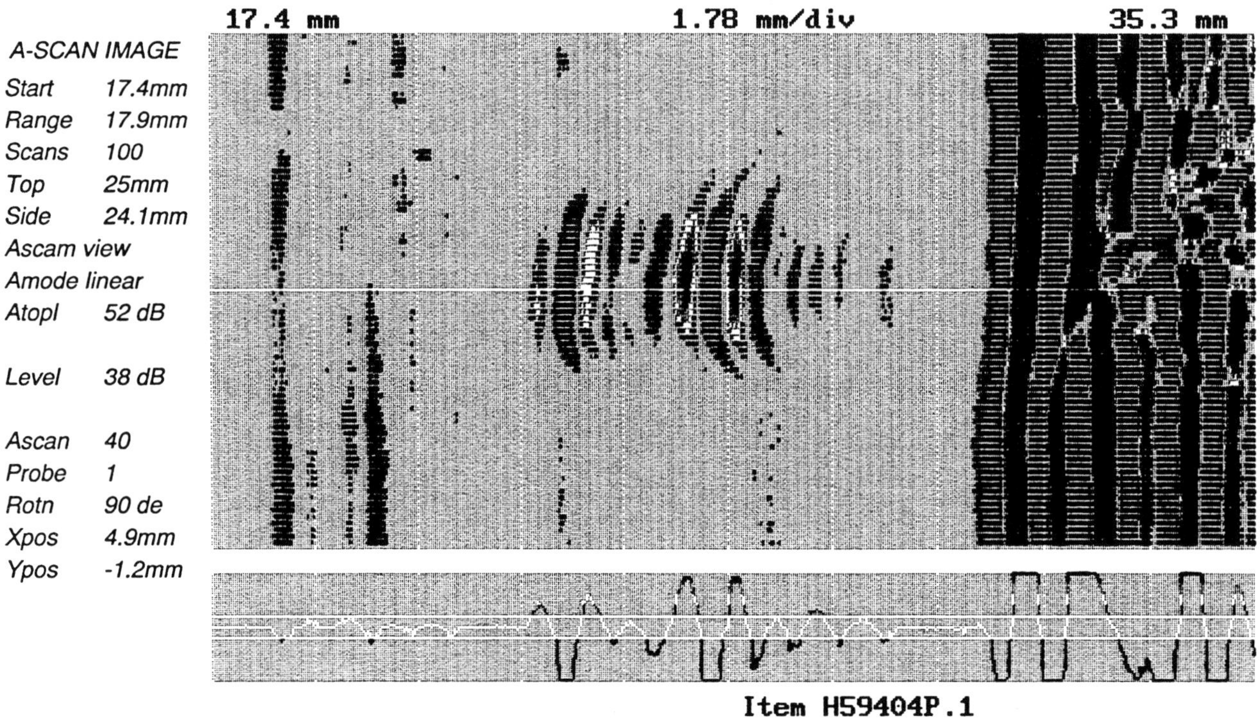

Fig.24 TOFD image of 4mm diameter foil insert.

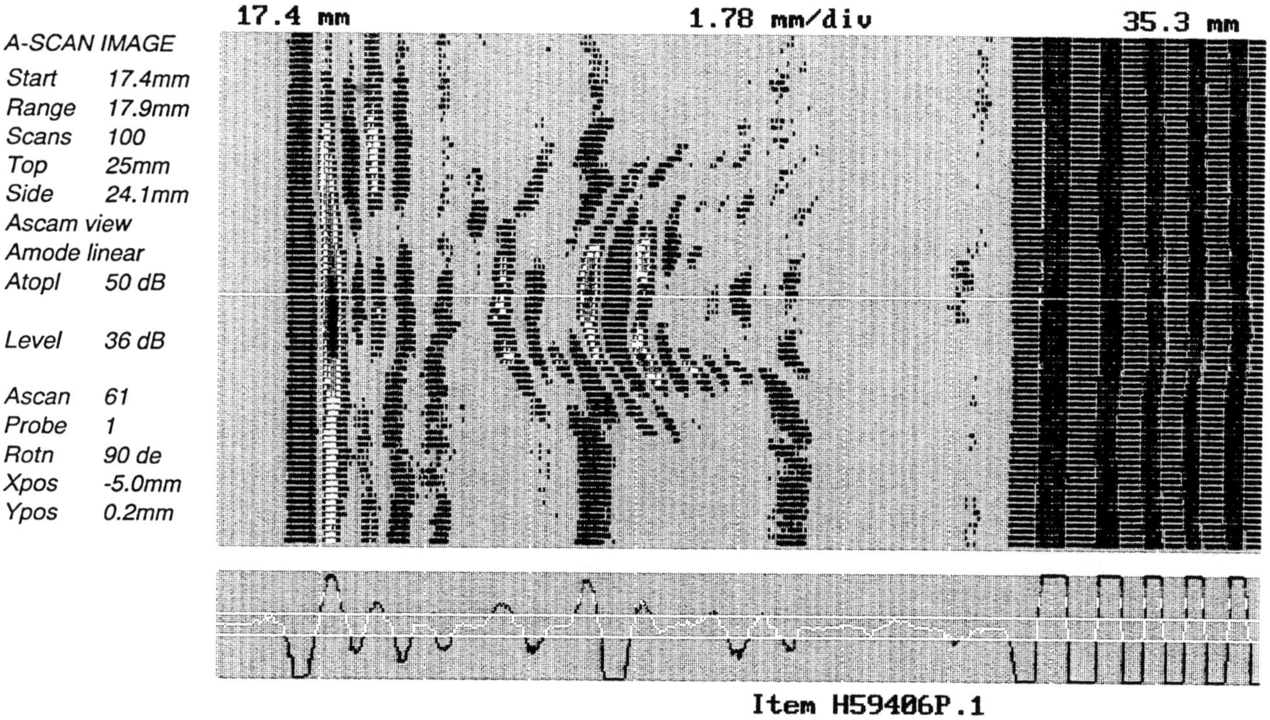

Fig.25 TOFD image of 6mm diameter foil insert.

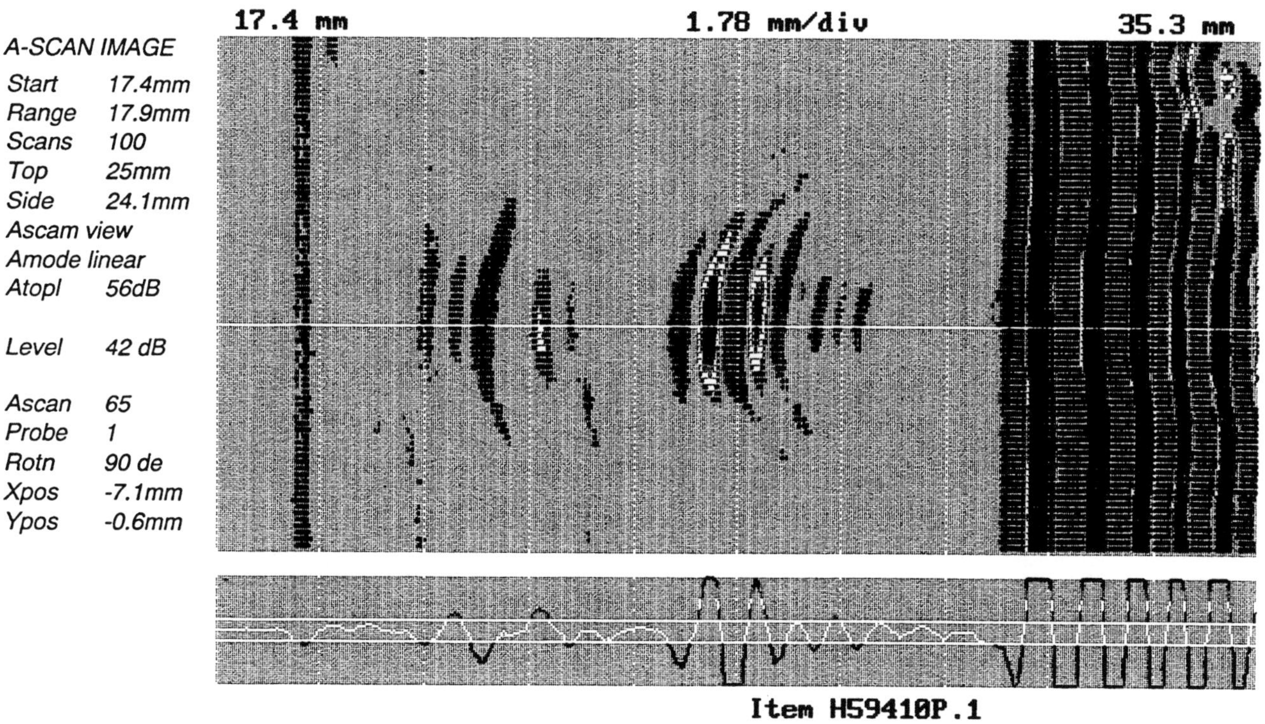

Fig.26 TOFD image of 10mm diameter foil insert.

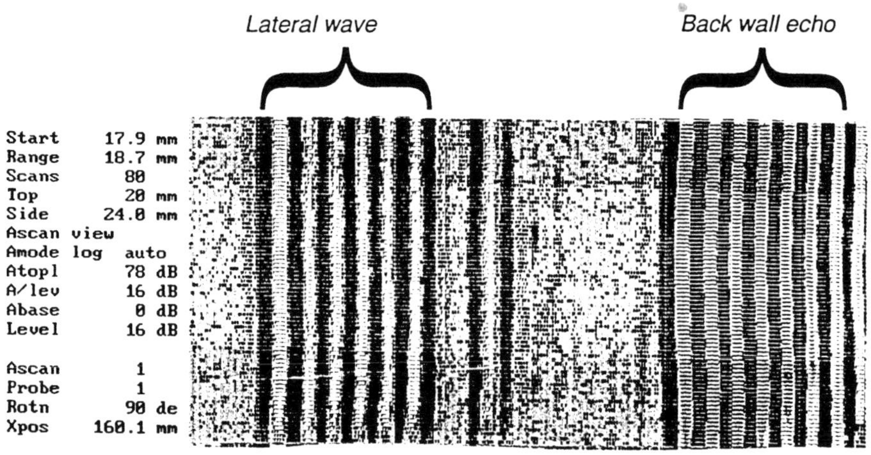

(a)

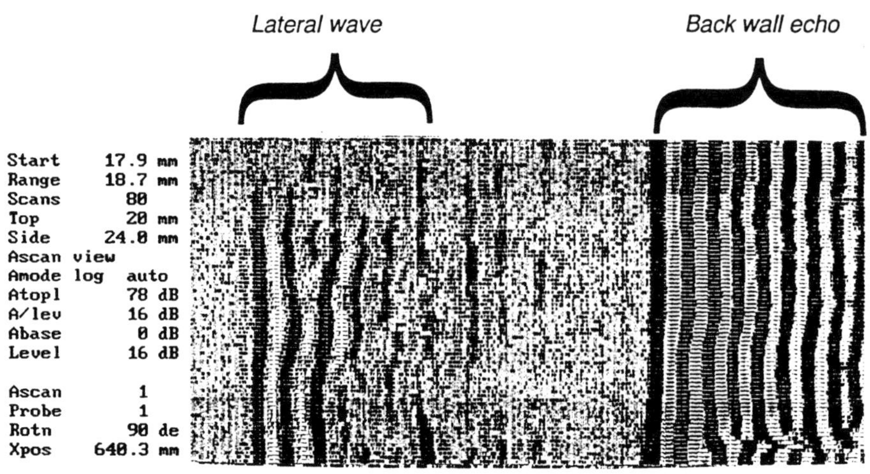

(b)

Fig.27 TOFD images from specimen H59-5: a) Dust free area; b) Area containing chalk dust

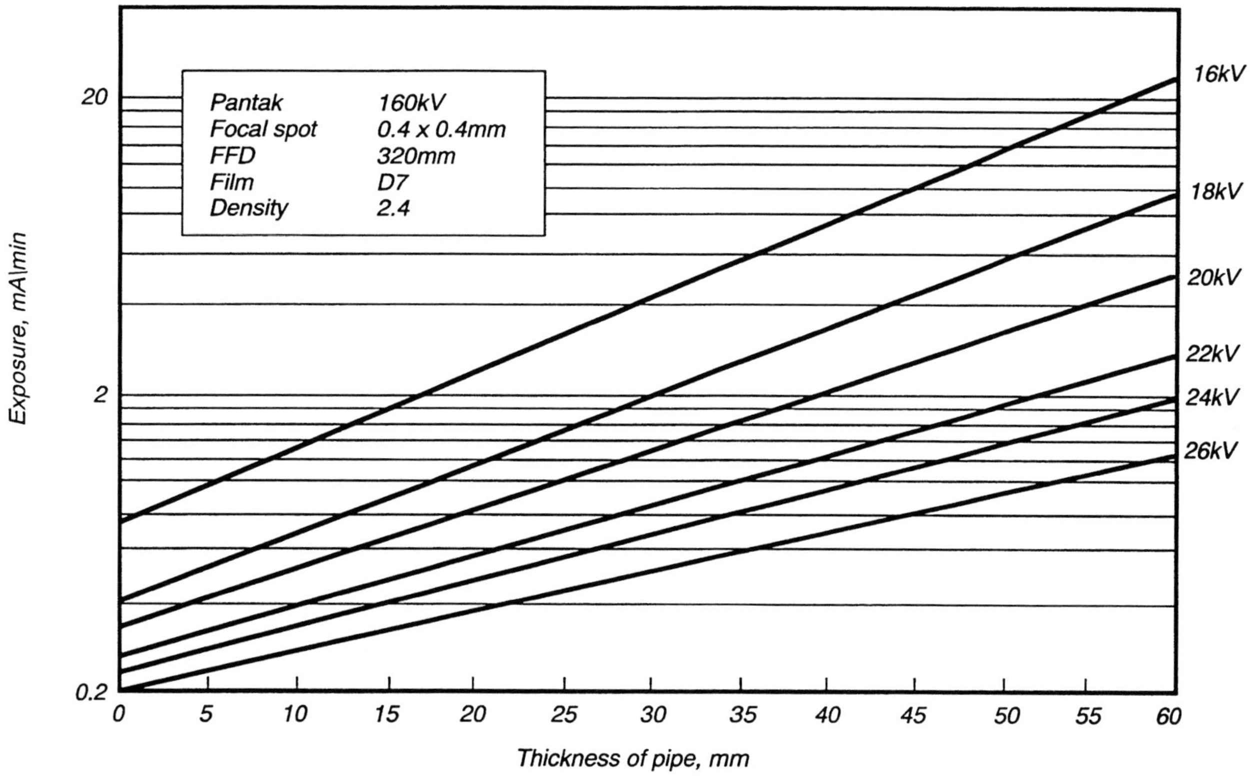

Fig.28 Exposure charts for group (a) polyethylene materials; Dowlex 2343E, Finathene 3802 B, Eltex TU B 121, Eltex TU B 124.

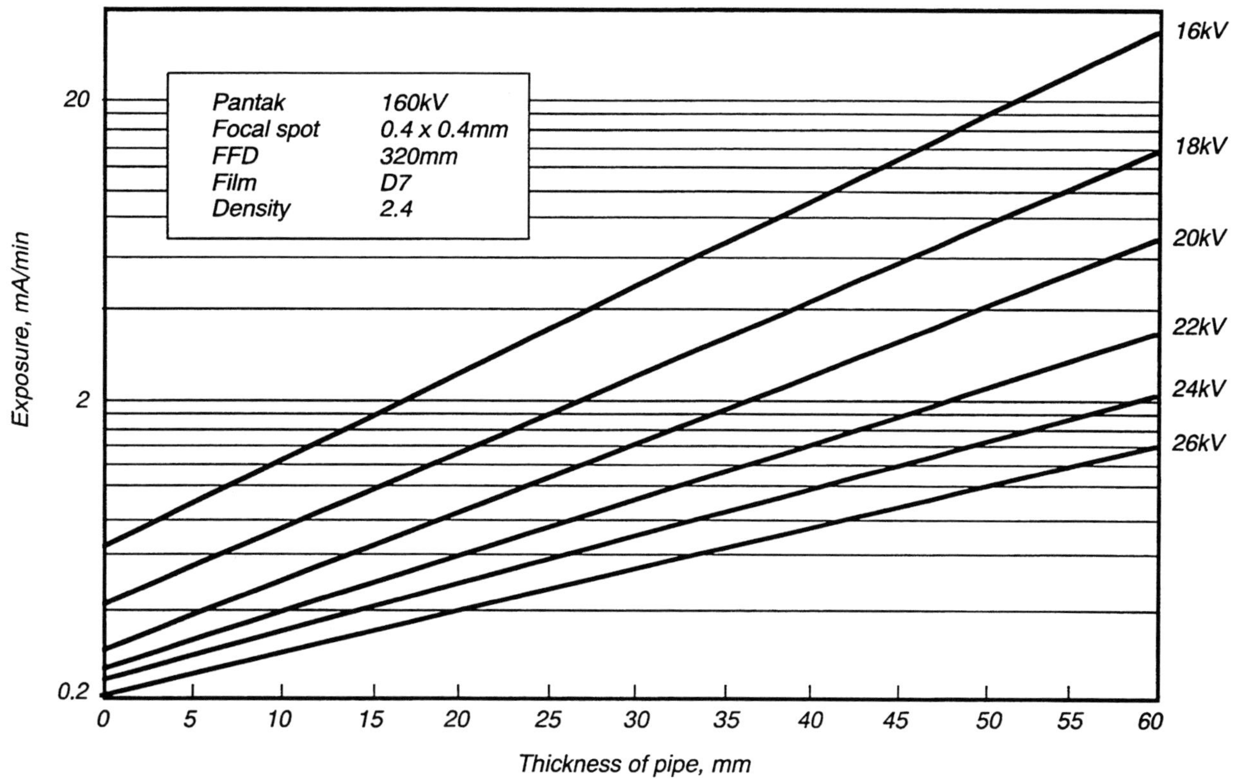

Fig.29 Exposure charts for group (b) polyethylene materials; Finathene 3802,Finathene 3802 Y, Eltex TU B 125.

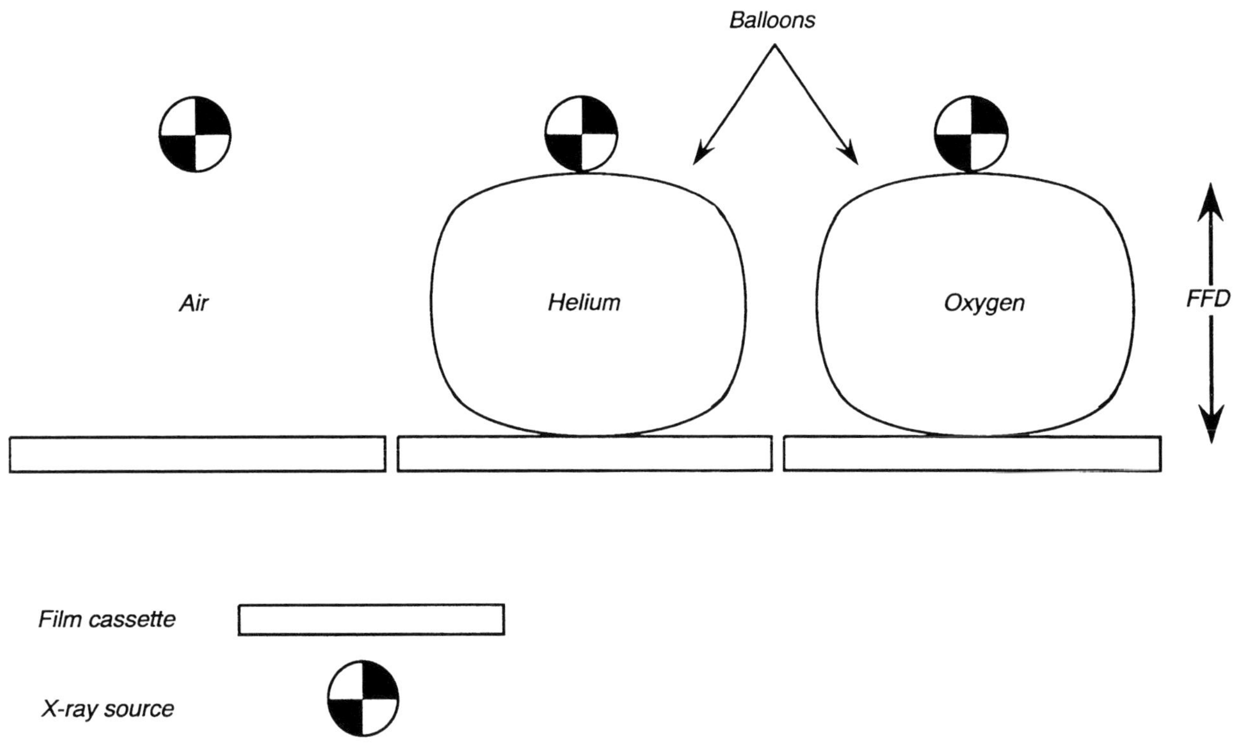

Fig.30 Experimental set up to compare X-ray absorption for air, helium and oxygen.

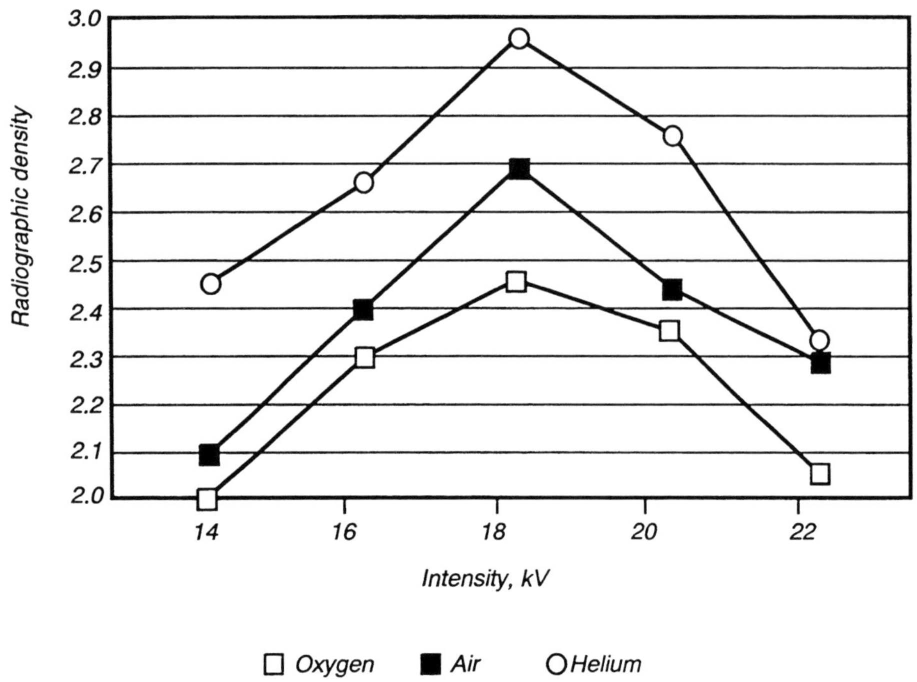

Fig.31 A comparison of the radiographic densities using air, helium and oxygen at low intensities.

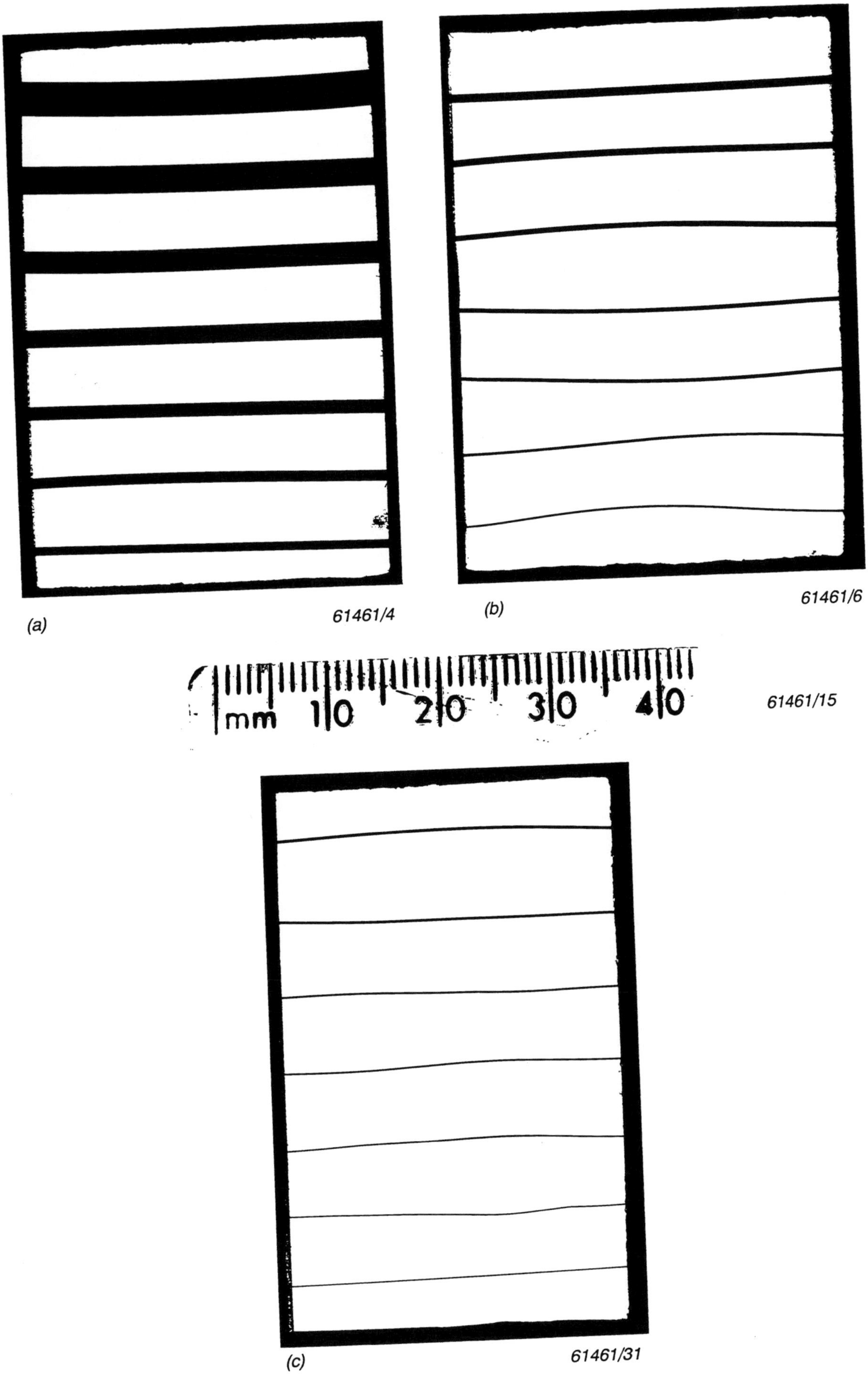

Fig.32 Photograph of polyethylene wire Type IQI: a) 0.8, 1.0, 1.25, 1.6, 2.0, 2.5, 3.2; b) 0.2, 0.25, 0.32, 0.4, 0.5, 0.63, 0.8; c) 0.63, 0.08, 0.1, 0.125, 0.16, 0.2, 0.25.

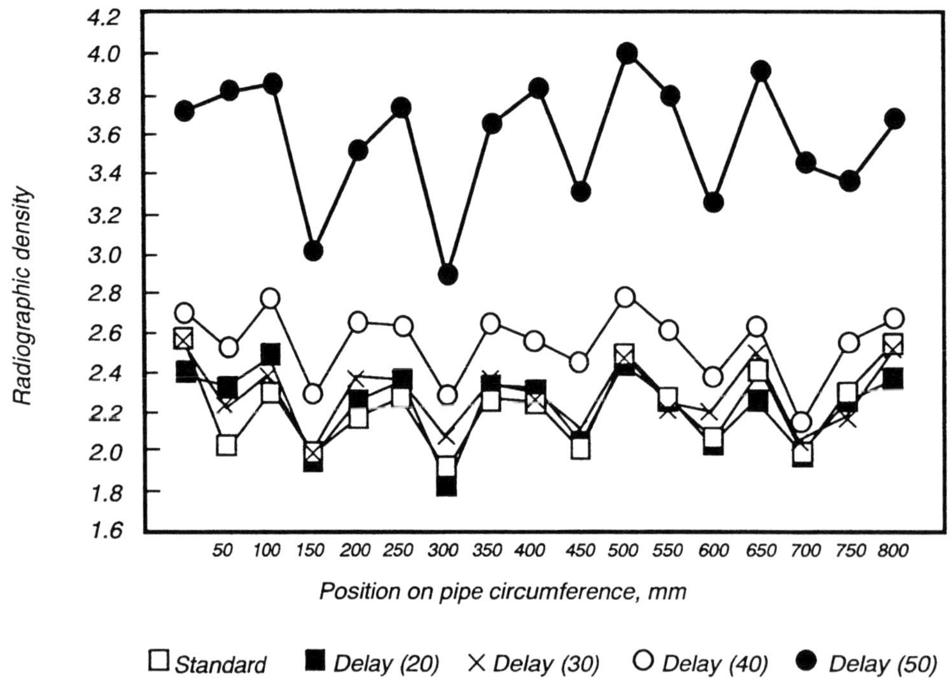

Fig.33 A comparison of radiograph density versus position around the pipe circumference, for: a) a standard weld; b) a weld with 20 second plate removal time; c) a weld with 30 second plate removal time; d) a weld with 40 second plate removal time; e) a weld with 50 second plate removal time.

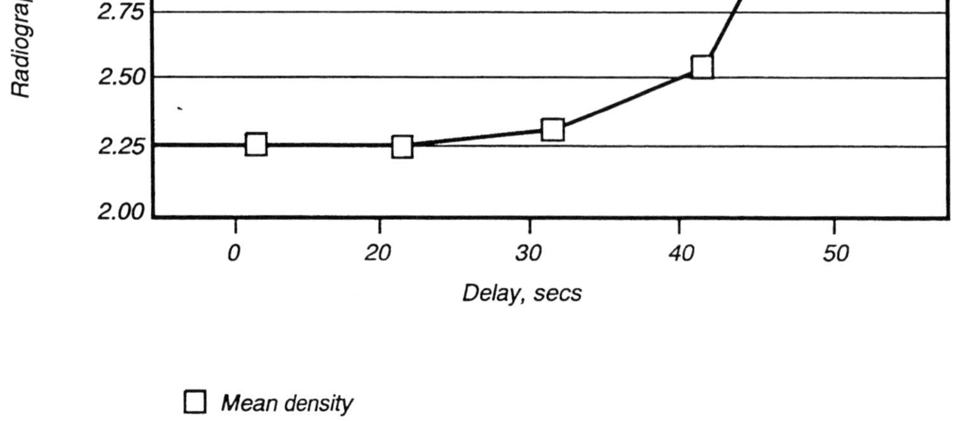

Fig.34 Mean radiograph density versus plate removal time.

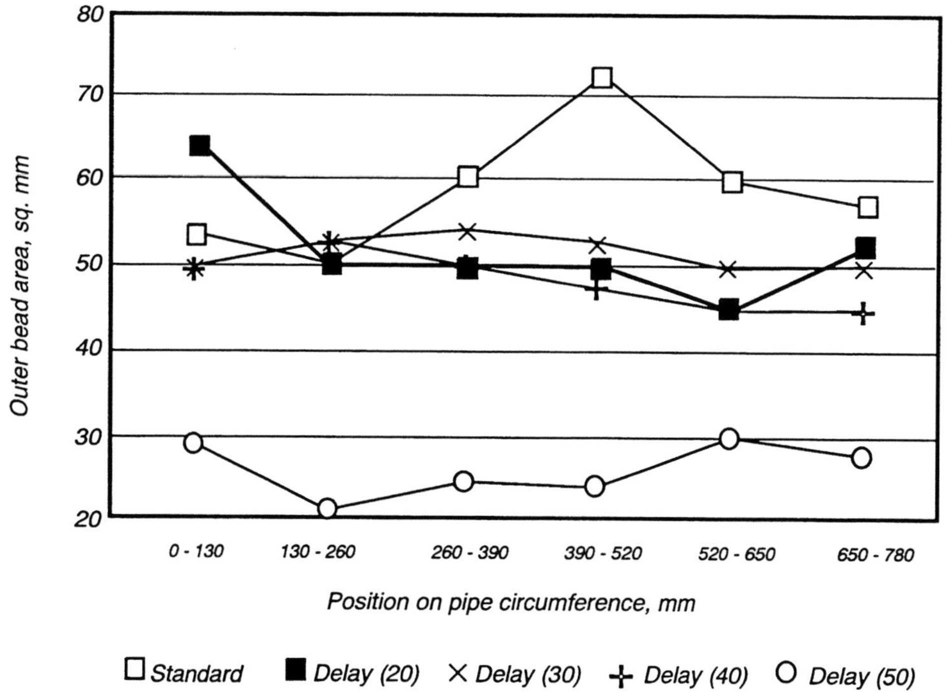

Fig.35 A comparison of the external bead cross-sectional area versus position around the pipe circumference, for: a) a standard weld; b) a weld with 20 second plate removal time; c) a weld with 30 second plate removal time; d) a weld with 40 second plate removal time; e) a weld with 50 second plate removal time.

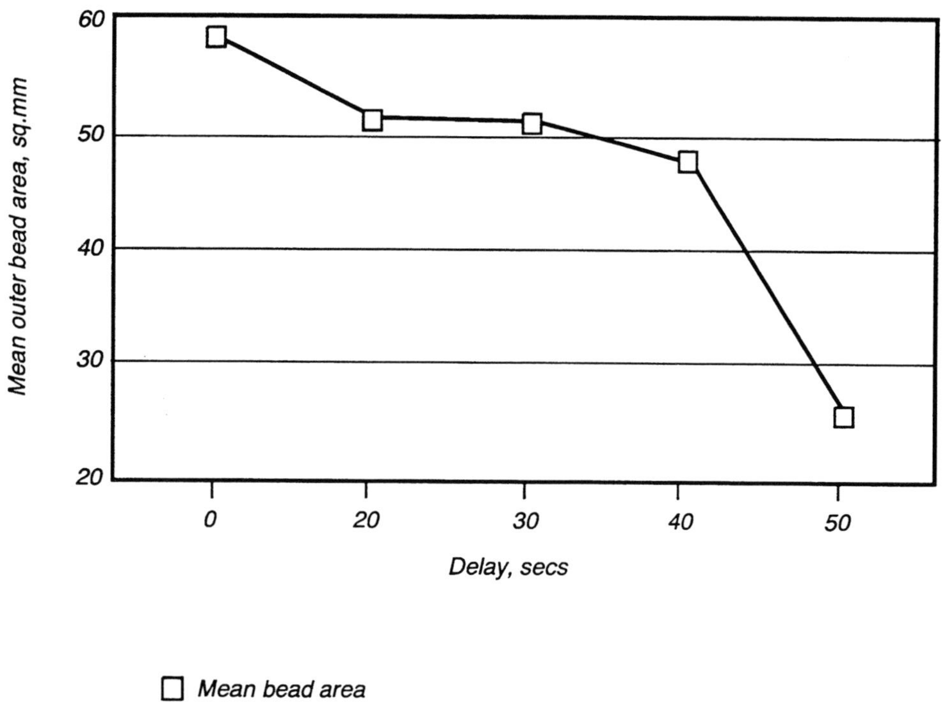

Fig.36 Mean external bead cross -sectional area versus plate removal time.

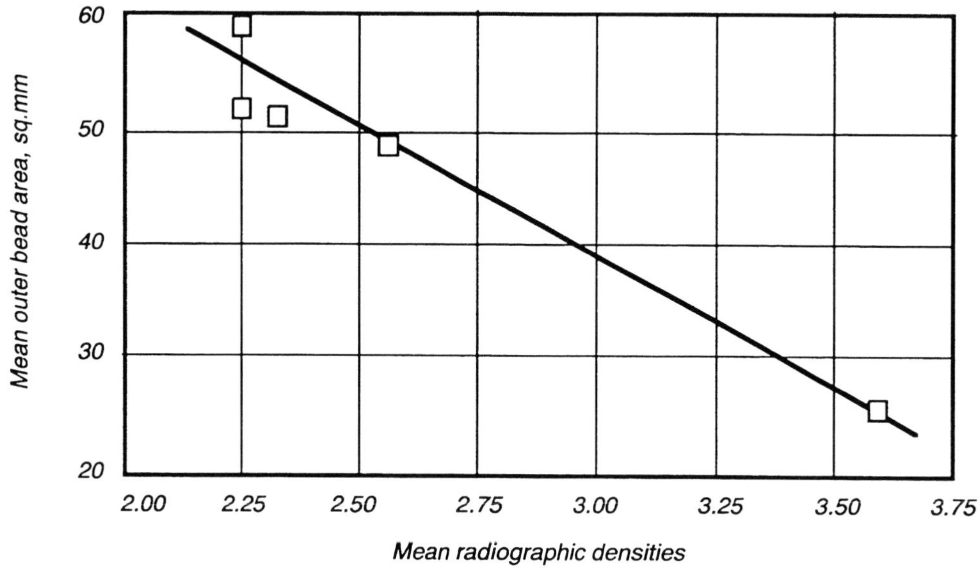

Fig.37 Mean radiograph density versus the mean external bead cross-sectional area.

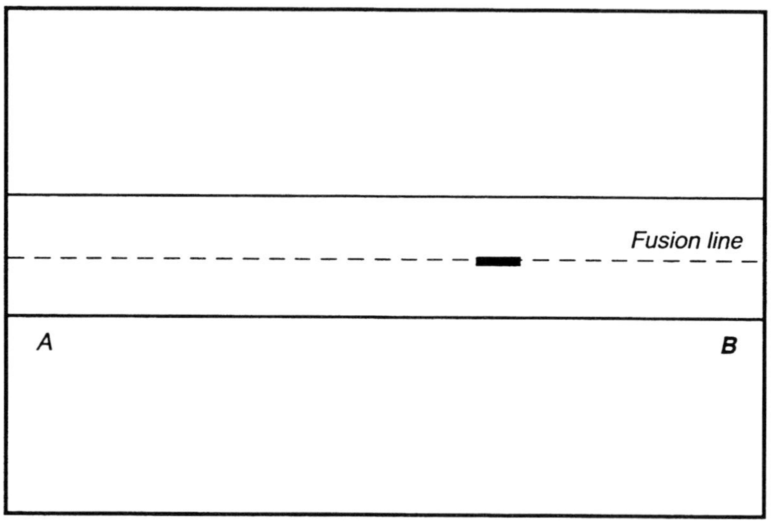

(a)

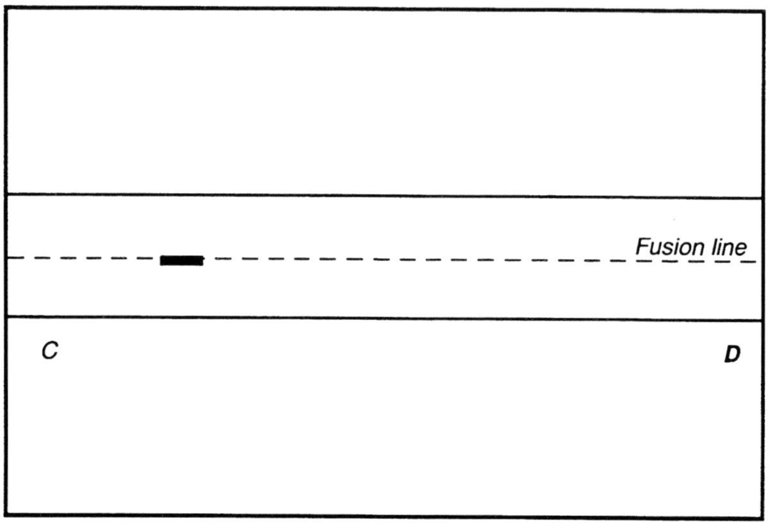

(b)

Fig.38 Illustrations of the weld made with a 30 second plate removal time (7075/43) with both beads on, showing indications a) 1 and b) 2.

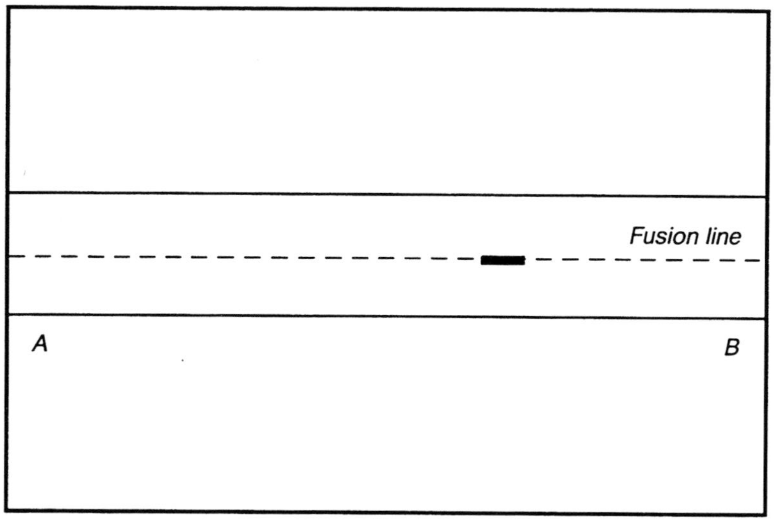

(a)

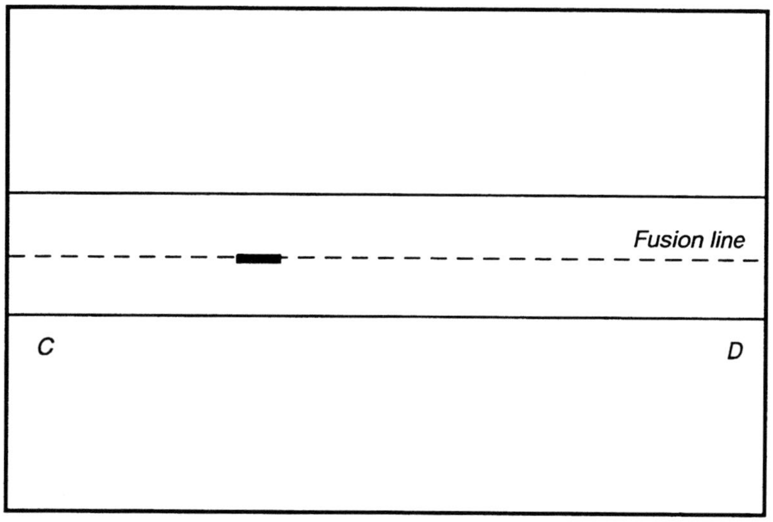

(b)

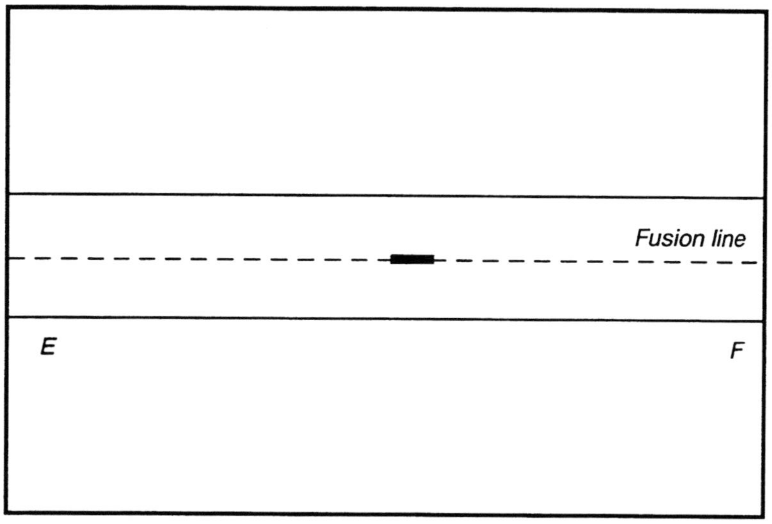

(c)

Fig.39 Illustrations of the weld made with a 30 second plate removal time (7075/43) outer bead removed, showing indications a)1, b)2 and c) 3.

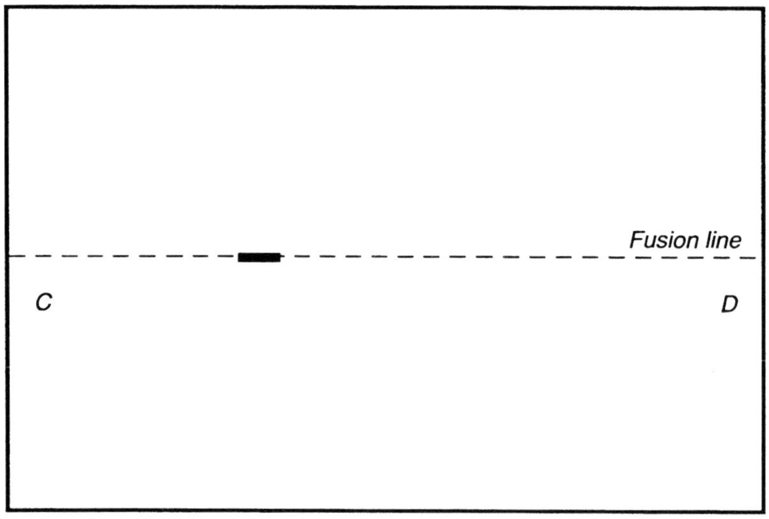

(a)

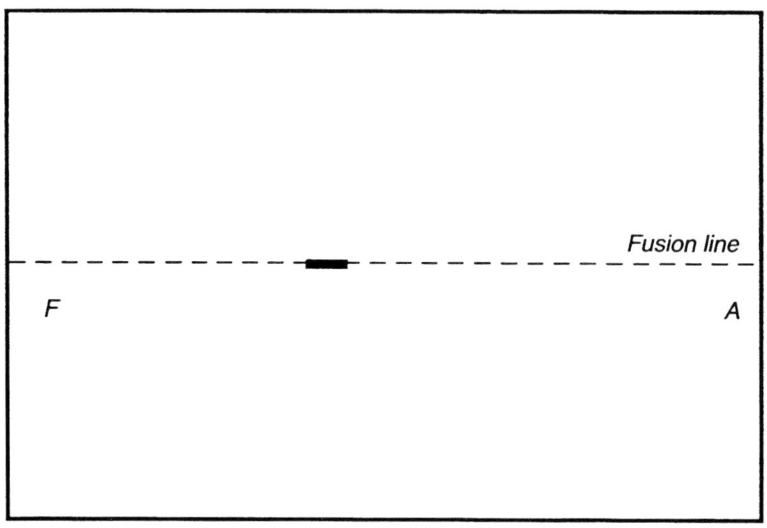

(b)

Fig.40 Illustrations of the weld made with a 30 second plate removal time (7075/43) with both beads removed, showing indications a)2 and b)4.

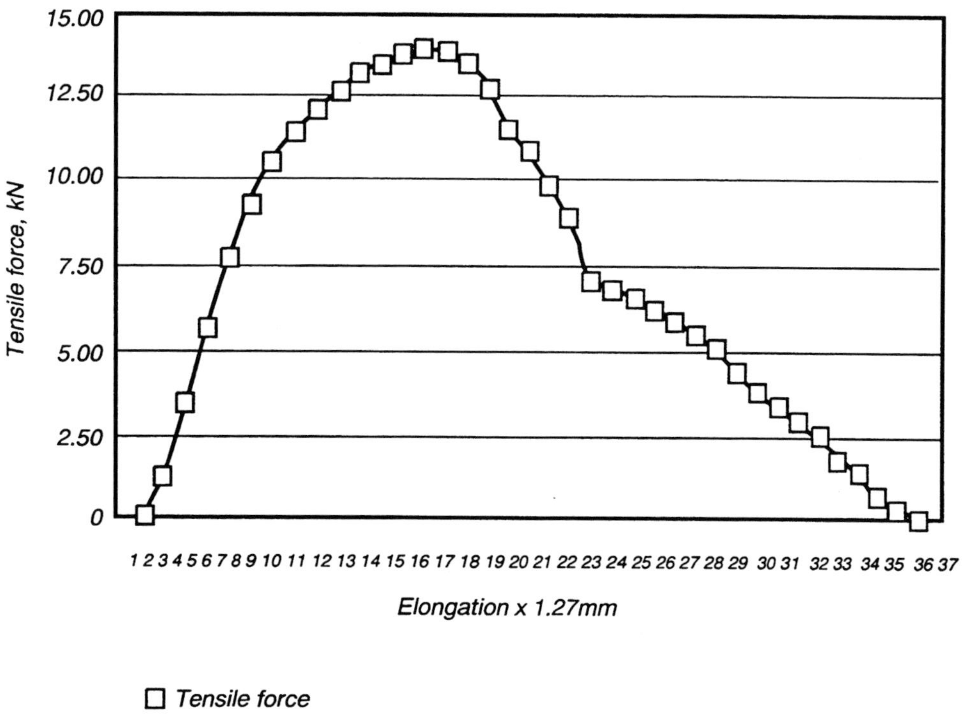

Fig.41 Tensile test result for the weld made with a 30 second plate removal time (7075/43).

APPENDIX A - Prototype Procedure For The Ultrasonic Examination of Fusion Welded Circumferential Butt Joints In Polyethylene Pipes

---

a) Scope

This procedure outlines various methods for the ultrasonic inspection of hot plate butt fusion welds in polyethylene pipes. The ultrasonic techniques employed include compression wave pulse echo, creeping wave pulse-echo, tandem and time-of-flight diffraction (TOFD).

b) Equipment

Flaw detectors - Krautkramer Branson USD10 (or equivalent
A-scan display) with minimum damping
Force Institutes PSP3 P-scan system with
MWS-2 scanner
Panametrics pre-amp
IBM PC (or equivalent)
PC-PROG program

Printers - Kodak Diconix 150 (for USD10)
Hewlett Packard Paintjet (for P-Scan)

Probes - 2 off Panametrics 1MHz A402S 12.7mm diameter transducers or equivalent
2 off Panametrics 2.25MHz A404S 12.7mm diameter transducers or equivalent
2 off 30° polytetrafluoroethylene (PTFE) wedges (approx. 60° compression wave in polyethylene (PE))
1 off 36° PTFE wedge (creeping wave in PE)
2 off 45° compression wave PE wedges
optional - 30,35,40 and 60° compression wave PE wedges
- all wedges to fit above transducers
Alternative probes if the above are not available:
60° shear wave probe (in steel) gives approx. 38° compression wave in PE
70° shear wave probe (in steel) gives approx. 45° compression wave in PE
(2 off) 80° shear wave probe (in steel) gives approx. 53° compression wave in PE
- all 2MHz if possible.

Testblocks - BS 2704:1978 A4 style block in polyethylene (no. 127)
BS 3923 Part 1:1986 3mm diameter side drilled hole DAC block in polyethylene (no. 116)
PE pipe section containing 5 and 10mm diameter end-drilled flat bottomed holes (no. 118)
PE block containing a range of notches (no. 85)

c) Calibration

Before inspection commences, a performance check must be carried out on the flaw detector in accordance with BS 4331 Part 1: 1978.

Calibrate the A-scan flaw detector for polyethylene using a 1MHz transducer on the side of the A4 block. Set the range of the flaw detector to 50mm. Echoes appear at 20mm and 40mm when the flaw detector is correctly calibrated. If possible, measure the thickness of the pipe to be inspected with a vernier and check that the velocity obtained is approximately the same as that calculated on the A4 block. (A typical value for compression waves in polyethylene is 2300m/sec). Measure the attenuation in several places, using 1st and 2nd reflections. Couple the transducer to each different wedge that is to be used during the inspection, and check the beam angles against reflections from the side drilled holes on the DAC block.

d) Pulse Echo Creeping Wave Technique

The creeping wave technique is designed to detect lack of fusion near the top of the weld.

Couple the 1MHz transducer to the 36° PTFE (creeping wave) wedge and maximise the signal from the 3mm deep notch in testblock no. 85. Adjust the delay on the A-scan flaw detector so that the range of the signal corresponds to the distance between the notch and the index point of the probe.

Measure the distance from the edge to the centre of the pipe weld bead. Place the creeping wave probe at this distance from the 1mm deep notch in testblock no. 85 and adjust the signal to 80% screen height. Couple the probe to the pipe to be inspected and place the probe up against the weld bead. Starting at a known datum point, move the probe around the circumference of the pipe using plenty of couplant. Observe the signal from the other side of the bead and check that this does not disappear. If it does, check the coupling first and re-scan. Note the position and amplitude of any indications greater than 16% screen height appearing in front of the signal from the bead as well as where the bead signal disappears. If the printer is attached, take a hard copy of the screen for each indication.

e) Pulse Echo Compression Wave Technique

This technique is to inspect the root area of the weld.

Couple a 2.25MHz transducer to a 45° polyethylene wedge and calibrate the A-scan flaw detector screen to 50mm range using the polyethylene A4 testblock.

Scan over the 3mm diameter holes in the DAC block and maximise the signal from the hole with the range nearest to that of the root bead. Set the signal to 80% full screen height. Scan around the circumference of the pipe and note any indications above 16% screen height.

This can also be carried out using a 2MHz 70° shear wave probe (in steel) which generates a 45° compression wave in polyethylene.

f) Tandem Technique

The tandem probe technique is used to inspect the through-thickness of the weld.

Couple each 1MHz transducer to a 30° PTFE wedge. (Alternatively, two 70° or 80° 2MHz shear wave probes for steel may be used). Secure each probe, facing towards the weld, one behind the other, in a jig to maintain a constant separation.

* *A-scan flaw detector*

This technique can be applied manually with the A-scan flaw detector.

Set the flaw detector to twin/dual probe and make the front probe the transmitter. Place the tandem set-up on the section of pipe containing the 5mm diameter flat bottomed holes and maximise the signal from the hole in the mid-wall position by adjusting the lateral position of the receiving transducer. Adjust the range of the signal so that it appears centrally on the screen and set the gain of the signal to 80% full screen height. Scan around the circumference of the pipe with the transmitting transducer positioned up against the weld bead and note any indications above 16% screen height occurring before the constant internal bead signal.

* *P-scan system*

For greater clarity the P-scan can be used to collect an image from the tandem set-up.

Strap the MWS-2 manual pipe scanner to the pipe and fix the tandem jig to the Y-module of the scanner. The P-scan should be in A-scan mode with the appropriate dataset entered in. Collect up to 80 A-scans per part of weld inspected. Save the data files onto a 3.5in. diskette and take a backup copy. Using the P-scan post processing package, PC-PROG, installed on an IBM PC (or equivalent), it is possible to display a composite A-scan image of each set of 80 A-scans collected. A colour printout of each composite A-scan image can be obtained if a suitable printer is attached.

Indications from perpendicular flaws may also cause a reduction/loss of the internal bead signal.

g) TOFD

Time-of-flight diffraction can be used for the detection and possible sizing of defects located in the through thickness of the pipewall.

Couple both 2.25MHz transducers to the 30° PTFE wedges. Secure each probe in a jig, with each probe facing the other across the weld bead. Attach the jig to the MWS-2 scanner.

Connect each probe to the P-scan system ensuring that the output from the receiving probe is routed through the Panametrics pre-amplifier and that the P-scan itself is set to TOFD mode. With an appropriate dataset installed, position the transmitter probe at a marked datum on the pipe. Adjust the display so that both the lateral wave and backwall echo are clearly visible. Collect up to 80 A-scans per part of weld inspected. Save the data files onto a 3.5in diskette and take a backup copy. Using the P-scan post processing package, PC-PROG, installed on an IBM PC (or equivalent), it is possible to display a TOFD image of each set of 80 A-scans collected. A colour printout of each TOFD image can be obtained if a suitable printer is attached.

Defect size can be measured from the display, provided both top and bottom signals are discernable.

Alternatively, the A-scan flaw detector can be used in RF mode to carry out a manual TOFD scan. In addition, two 2MHz 80° shear wave probes may be used in place of the above as they become 53° compression wave probes in PE.

h) Evaluation And Reporting

All indications above 20% of the reference reflector should be noted for position and amplitude. Defect length should be measured by the 6dB drop method or the maximum amplitude method. A drawing should be made showing any indications reported.

i) Acceptance Criteria

At present, no acceptance criteria exist.

j) Personnel Certification

Operators must hold a current PCN Level 2 (or equivalent) in ultrasonics and have practical testing experience, including the use of TOFD and the P-scan system.

k) Associated Documents

BS 2704:1978 Calibration blocks for use in ultrasonic flaw detection.

BS 3923 Part 1:1986 Ultrasonic examination of welds.

BS 4331 Part 1:1978 Assessing the performance characteristics of ultrasonic flaw detection equipment.

BS 7706:1993 Guide to Calibration and setting–up of the ultrasonic time–of–flight diffraction (TOFD) technique for the detection, location and sizing of flaws.

G.A. Georgiou and I.A. Macdonald 'Ultrasonic and radiographic NDT of butt fusion joints in polyethylene' TWI Members Report 465/1993.

APPENDIX B – Prototype Procedures For Radiographic Examination Of Fusion Welded Circumferential Butt Joints In Polyethylene Pipes

It is intended to rewrite these prototype procedures in a formal format as a proposed BSI standard.

a) Scope

These radiographic inspection procedures are intended for quality assurance of hot plate butt fusion welds in polyethylene pipes. The pipes covered are manufactured from polyethylene resins specified in Table B1 of this Appendix. The range of thicknesses covered is from 5–50mm, which includes both the inner and outer weld beads. These procedures do not deal with acceptance/rejection criteria.

b) Equipment

X–ray radiography set:	Pantak 160kV with beryllium windows and focal spot 0.4 x 0.4mm (or equivalent);
Film:	Paper cassette type, D7 speed, no screens no filters;
IQIs:	Either wire Type I made from Finathene 3820 B (for group (a) polyethylenes) or 3820 Y (for group (b) polyethylenes) according to the specifications in BS EN 462–1.

c) Personnel

Radiographic inspection and interpretation shall be supervised by a PCN approved radiologist.

d) Technique

i. Sample Identification and Makers

Any necessary pipe sample identifications and markers will be marked with permanent ink, along with corresponding lead markers attached with clear adhesive tape to the pipe at appropriate positions. These include:

Sample identification (alphanumeric);
Primary datum point (1.5cm from the weld);
Direction of rotation;
Further datum points (as necessary).

ii. Focus–to–Film Distance, Exposure Charts and Density

The focus–to–film distance will be 320mm. One of two exposure charts will be used according to whether the polyethylene material falls into group (a) or (b). For group (a) (Dowlex 2343 E, Finathene 3802 B, Eltex TU B 121, Eltex TU B 124) use exposure chart for Finathene 3802 B (Fig.B1 in this Appendix). For group (b) (Finathene 3802, Finathene 3802 Y, Eltex TU B 125) use exposure chart for Finathene 3802 Y (Fig.B2 in this Appendix).

The density of the radiographs shall be between 1.8 and 3.0 over the area of interest of each radiograph.

iii. The Straight Technique

The straight technique is when the source is in the weld plane (see Section (e), Georgiou and Macdonald) and a double wall single image technique shall be used. The number of radiographs shall be in accordance with BS 2910. Separate radiographs of the welded joint shall overlap sufficiently to ensure no portion of the joint remains unexamined. The sample shall be placed on a lead platform when being radiographed to reduce backscatter. This is the primary technique for detecting and sizing flaws.

iv. The Throwing Technique

The throwing technique is when the source is offset from the weld plane (see Section (e), Georgiou and Macdonald). The remaining parameter specifications are as for the straight technique above. This is the secondary technique and shall be used when additional information is required regarding the character of the flaw (e.g. shape, a gauge of through-thickness height).

Up to two offset positions may be considered, equivalent to 15 and 20° from the vertical respectively.

v. Viewing, Interpretation and Reporting

Viewing shall be performed under conditions specified in this report. The IQI sensitivity shall be determined in accordance with Table B2 in this Appendix. Radiographs of welds under standard conditions and ones with manufactured flaws may be used to assist interpretation. Details of any flaws such as inclusions, voids or nominally cold welds shall be reported on TWI standard radiographic report sheets, together with reference to this prototype procedure.

e) Associated Documents

BS EN 462-1: 1994: 'Non-destructive testing - Image quality of radiographs'. Part 1: Image quality indicators (wire type) - Determination of image quality value.

BS 2910:1986: 'Radiographic examination of fusion welded circumferential butt joints in steel pipes'.

Georgiou G A and Macdonald I A: 'Ultrasonic and radiographic NDT of butt fusion joints in polyethylene'. TWI Members Report 465/1993.

Table B1 Product specification of medium to high density polyethylene materials used in the study

** Product	Colour	Density (g/cm^3)	Yield strength (N/mm^2)	Modulus (N/mm^2)	Hardness Shore D
Dowlex	White	0.937	17	710	60
Finathene 3802 B	Black	0.948	19	700	59
Eltex TU B 121	Black	0.958	25	1200	59
Eltex TU B 124	Dark blue	0.951	25	1200	60
* Finathene 3802	Light blue				
Finathene 3802 Y	Yellow	0.941	19	700	59
Eltex TU B 125	Orange	0.951	25	1200	59

* Details not supplied by manufacturer
** These are the Trade Names of the products as supplied by Stewarts and Lloyds Plastics

Table B2 IQI sensitivity for critical and normal techniques

Specimen thickness, mm	Section A Critical techniques	Section B Normal techniques
	Minimum no. of wires visible	Minimum no. of wires visible
5	7	6
10	5	4
15	5	4
20	5	4
25	5	4
30	5	4
35	5	4
40	4	3
45	4	3
50	4	3

ASSESSMENT OF AGEING PROPERTIES AND RESIDUAL STRESSES IN THERMOPLASTIC WELDS

By: **S M Tavakoli and R H Leggatt**

Background

Polymeric materials are susceptible to degradation during processing or in service. This can lead to deterioration in properties and premature failure. During welding polymers are exposed to elevated temperatures and considerable mechanical forces are applied to the viscous polymer. This process can generate thermal and residual stresses and initiate degradation reactions thus affecting the long-term performance of the welded components. There is no established technique available for measuring residual stresses in thermoplastic welds. Therefore it became necessary to develop a method for determining residual stresses in welds

Objective

1. To develop a technique for the measurement of residual stresses at welds in polymers.

2. To generate data on residual stresses at welds in PEEK, PES, PEI, ABS and GR polyester.

3. To determine the effect of ageing on residual stresses, mechanical properties and toughness in PEEK.

Experimental Approach

Five thermoplastic welds and parent materials were selected for measurement of residual stresses. These included polyetheretherketone (PEEK), polyetherimide (PEI), glass-reinforced (GR) polyester, polyethersulphone (PES) and acrylonitrile-butadiene-styrene (ABS). A satisfactory condition was established for hot plate welding of PEEK. PEEK welds and parent materials were aged in air at 250°C for 140 days and in boiling water for up to 28 days. The effects of ageing on tensile properties of PEEK were investigated.

A two-stage sectioning technique was developed for measuring the distribution across the weld of residual stresses parallel to the welding direction in welded polymer specimens. This involved cutting out a 20mm wide strain gauged strip containing the weld, and then machining off 2mm wide layers from the edge of the strip.

The effects of ageing in air at 250°C and after immersion in boiling water were determined at different ageing times.

Results and Discussion

The experimental procedure developed in this project was found to be suitable for measuring the distribution of longitudinal residual stresses at welds in polymers with an estimated accuracy of +0, −1.5N/mm^2. The maximum stresses were found at the weld centreline, and were 15N/mm^2 in PEEK 5 months after welding, decreasing to 12N/mm^2 12 months after welding, and then remaining stable. Residual stresses in PEEK were unchanged after 72 hours in boiling water, but decreased to zero after 72 hours in air at 250°C. The maximum stresses in welded GR polyester, PEI and ABS specimens were 23, 16 and 6N/mm^2 respectively (see Fig.12). In each material, the maximum residual stresses were less than 20% of the tensile strength.

During ageing of PEEK in air at 250°C, there was an increase of 18% in yield (σ_y) and a 27% increase in tensile strength at break (σ_b) when ageing time was increased from 14 to 140 days. The elongation at break decreased from 6.6mm to 1.8mm after the same period of ageing. In general the tensile properties of the parent materials were less affected by ageing than was the case for the welds.

Immersion in boiling water for 21 days did not cause any noticeable changes in tensile properties of PEEK welds or parent materials.

Main Conclusions

- A procedure was developed to measure residual stresses at the weld.

- The maximum longitudinal residual stresses at welds in PEEK were:

  a) 15N/mm^2, aged in air for 5 months

  b) 12N/mm^2, aged in air for 12 months with no further reductions for specimens aged for 21 and 28 months.

- The maximum stresses in welded GR polyester, PEI and ABS were 23, 16 and 6N/mm^2 respectively.

- Ageing of tensile specimens of PEEK in air at 250°C showed that welds degraded faster than the parent materials.

Recommendations

The measurement procedures and residual stress data described in this report may be used for investigations of the effect of residual stresses on degradation mechanisms and structural integrity of welded polymer components.

ASSESSMENT OF AGEING PROPERTIES AND RESIDUAL STRESSES IN THERMOPLASTIC WELDS

By: S M Tavakoli and R H Leggatt

1. INTRODUCTION

Polymeric materials are susceptible to degradation when exposed to elevated temperatures, light, chemical agents, stresses or other forms of energy, during processing or in service [1–4]. This can lead to deterioration in properties and premature failure.

Fabrication of polymers into finished products often requires melting. During processing operations, such as injection or transfer moulding, extrusion or welding, considerable mechanical forces are applied to the viscous polymer melt causing the polymer chains to undergo homolytic scission at some carbon–carbon bonds to produce macroalkyl radicals. These reactive chemical species are very active and can initiate degradation reactions and influence the long–term performance of the materials.

TWI has been conducting a programme of work to study effects of welding and ageing on a number of engineering thermoplastics. Effects of ageing in air at 100, 110, 120°C and immersion in warm (80°C) and boiling water on the tensile properties of hot plate welds in ABS were studied [5,6]. Ageing adversely affected tensile properties of the weld more than the parent material. The weld cross–sectional area was also considerably changed during ageing. Fourier transform infrared microspectroscopy (FTIR) indicated that chemical changes associated with ageing were more significant in welds than in the parent material. Dynamic mechanical thermal analysis (DMTA) of ABS welds showed that the damping peak (tan δ) of polybutadiene shifted to higher temperatures and decreased in height on ageing in air at 110°C.

In a recent study, effects of ageing in air at 220 and 230°C and immersion in boiling water on the tensile properties of hot plate and linear friction welds in both unreinforced and glass–fibre reinforced polyethersulphone (PES) were investigated [7,8]. Ageing had a more adverse effect on properties of the welds than parent materials. The parent materials and linear friction welds showed higher retentions of strength for the reinforced than the unreinforced PES grades when aged at 220°C, but this trend was reversed for the hot plate welds. Immersion in boiling water reduced the strengths of the reinforced PES, hot plate and linear friction welds more than the unreinforced grades. Fractographic examination showed that most tensile fractures initiated from cavities formed during ageing. Dynamic mechanical thermal analysis (DMTA) showed an increase in glass transition temperature (Tg) of the PES for the hot plate weld on ageing.

Polyetheretherketone (PEEK) is another important engineering thermoplastic suitable for high temperature applications in extreme environments, in many industries, including electronic, aerospace and automotive. PEEK is insoluble in all common solvents and has excellent resistance to a very wide range of organic and inorganic liquids. PEEK has an estimated continuous working temperature of 250°C. A brief literature review on degradation of PEEK is presented in Appendix 1.

It seems likely that the initial strength and long term structural integrity of welded joints in polymers are affected by the residual stresses generated during welding. Residual stresses contribute to the driving force for stress-dependent failure mechanisms such as brittle fracture and environmental stress cracking. Hence there is a need for the generation of data on residual stresses in polymers after welding and ageing.

The present authors are not aware of any previous measurements of residual stresses of welds in polymers. Various workers have used layer removal methods to measure the through-thickness variation of residual stresses due to manufacturing processes in polymer and composite materials, including injection moulded thermoplastics [9], injection moulded polycarbonate rectangular bars [10] and carbon fibre thermoplastic matrix laminates [11–13]. Sullivan and Matsuoka [14] described the use of a photoelastic technique to measure stress patterns at inserts in epoxy disks. Young et al [15] and Filiou et al [16] have used Raman spectroscopy to measure strains in the fibres in carbon fibre reinforced PEEK.

Of the measurement methods described above, photoelasticity is only suitable for transparent materials, and Raman spectroscopy measures the local strains in the reinforcement rather than bulk strains in the component. Hence it was decided to use the layering technique in the present study. However, it was necessary to modify the experimental procedure in order to measure the in-plane variation of residual stresses across the weld, rather than the through-thickness variation of stresses which have been measured in previous investigations [9–14].

2. OBJECTIVES

1. To develop a technique for the measurement of residual stresses at welds in polymers.

2. To generate data on residual stresses at welds in PEEK, PES, PEI, ABS and GR polyester.

3. To determine the effect of ageing on residual stresses, mechanical properties and toughness in PEEK.

3. PROGRAMME OF WORK

A. Preparation of hot plate welds of injection moulded PEEK plaques.

B. Ageing of tensile specimens produced from parent PEEK material and welded plaques in the following environments:

(i) Air at 250°C for up to 140 days.

(ii) Air at room temperature for 28 months.

(iii) Boiling water for 28 days.

C. Determination of mechanical properties of unwelded and welded PEEK for both unaged and aged conditions.

D. Assessment of residual stresses in unwelded and welded PEEK, PEI, GR polyester, ABS and PES, both unaged and aged.

4. EXPERIMENTAL PROCEDURE

4.1. Materials

The following materials were used for welding, ageing and assessment of residual stresses.

(i) Polyetheretherketone (PEEK) – Victrex 450G, a general purpose grade for extrusion or injection moulding, manufactured by Victrex Ltd.

(ii) Polyethersulphone (PES) – Victrex 4100G was the unreinforced grade and Victrex 4101 GL30 was used as the reinforced grade (glass, 30%), manufactured by ICI.

(iii) Acrylonitrile–butadiene–styrene (ABS), Terluran 967K manufactured by BASF.

(iv) Polyetherimide (PEI) – Ultem 1000, unreinforced PEI, manufactured by GE Plastics.

(v) Glass reinforced polyester – Valox 420 SEO, 30% glass reinforced (GR) thermoplastic polyester, manufactured by GE Plastics.

The identity of these materials was verified by Fourier infrared spectroscopy (FTIR) prior to welding.

4.2. Welding

4.2.1. Hot Plate Welding of PEEK

4.2.1.1. Welding Equipment

The welding trials were conducted on a small part hot plate welding machine. This machine is a commercial unit manufactured by Bielomatic Leuze GmbH (model HV 47602, TWI machine No. 04/P/3). A resistance heated aluminium bronze platen of 150 x 100mm was used. Due to the high plate temperatures involved it was not possible to use the usual non–stick PTFE coating on the plate. The main machine controls and their functions are listed in Table 1. The sequence of machine operations to complete a weld joint in a pair of channel section parts is shown schematically in Fig. 1.

4.2.1.2. Instrumentation

A six channel Watanabe Multicorder MC6601 chart recorder (TWI No. D130) was used to monitor weld cycle time and material displacement for all welding trials conducted in this programme. The displacement during heating and weld consolidation phases were taken from a Penny and Giles rectilinear potentiometer. The hot plate temperature was measured immediately prior to each weld. A temperature probe was held against the hot plate and a direct temperature reading taken after fifteen seconds. The heating and weld consolidation pressures were recorded from calibration graphs determined by mounting a load cell between dummy workpieces placed in the platens of the welding machine. The calibration graphs were derived from load cell readings against gauge pressure settings.

4.2.1.3. Materials and Specimen Form

Injection moulded plaques of PEEK were supplied by ICI. For the parameter investigation, the plaques were cut into eight, as shown in Fig. 2a. The remainder of the plaques was cut into four, as shown in Fig. 2b. All edges to be welded were machined flat. The edges to be welded were transverse to the flow direction of the material, as shown in Fig. 2.

Prior to welding the material was dried for 3 hours at 150°C in a vacuum oven.

4.2.1.4. Welding Procedure

Prior to positioning the components, the joint faces were wiped with MEK to remove any surface contamination. Following clamping, the specimens were brought into contact with each other to set the datum level for measurement of material displacement occurring during welding. The approach rate between the testpieces and the hot plate was maintained at 0.2m/sec for all the welding trials. The depth of melt stops was fixed at 0.6mm. The welding equipment was then used in automatic cycle to sequence the welding operation through the heating and weld consolidation stages. In between each weld, the plate was scraped with a knife and wiped with wire wool to ensure any resin that had adhered to the plate was removed.

The following range of parameters was investigated:

Heating Pressure	0.2 – 0.3N/mm^2
Heating Time	5 – 120 secs
Plate Temperature	350 – 400°C
Welding Pressure	0.2 – 0.3N/mm^2
Cooling Time	30 secs
Depth of Melt	0.6mm

4.2.1.5. Assessment

The effect of the welding conditions investigated was assessed through visual appearance, measurement of material displacement durign welding and tensile strength.

The tensile testing was carried out on an Avery Denison universal testing machine (TWI Machine No. 04/R/13) at a crosshead displacement of 5mm/min. The welded samples were machined into the required test specimen geometry, shown in Fig. 3a and b. Initially, attempts were made to remove the weld flash, but, due to the brittle nature of some of the joints, this was not practical. Except where specified, the joints were tested with the flash left on. The weld strength was calculated on the basis of the original cross sectional area of the testpiece interface.

4.2.2. Welding of ABS, PES, PEI and GR Polyester

The optimum conditions previously established for hot plate welding of ABS [5] and PES [7] were used.

Hot plate welding of PEI and GR polyester was carried out using an optimum condition [17].

4.3. Ageing Procedures for PEEK

4.3.1. Air oven ageing

A Wallace multi-cell ageing oven was used as described previously [5]. PEEK welds and parent materials were exposed to ageing in air at 250°C. The 250°C was selected as an ageing temperature because this is known as an estimated continuous working temperature for PEEK. The glass transition temperature (Tg) of PEEK 450G is 143°C and the melting point is 334°C. Specimens were removed after different time intervals of 14, 28, 56, 112 and 140 days. Tensile properties as well as weight and cross-sectional areas were determined for each specimen after a given period of exposure.

4.3.2. Water immersion tests

The effects of full immersion in tap water at boiling point on PEEK welds and parent materials were evaluated as described previously [5]. Specimens were exposed to ageing for up to 28 days and removed after 7, 14, 21 and 28 days for determination of tensile properties.

4.4. Mechanical Tests

4.4.1. Tensile testing of PEEK

For optimisation of welding conditions for PEEK, tensile testing was carried out as described in section 4.2.1.5.

Tensile properties of PEEK specimens (George Fischer geometry, See Fig.3) used in ageing studies were determined at a loading rate of 5mm/min. Samples were either in the unaged, as-welded condition or had been exposed to ageing in air at 250°C or boiling water. Welded samples were tested transverse to the weld, which was located at the midpoint of the gauge length of the specimens.

4.4.2. Fracture mechanics tests of PEEK

Single edge notched bend specimens (See Fig.4) were prepared from PEEK parent and welded material, with a sheet thickness (3.1mm) being used as the specimen thickness (t). 2t x t (6 x 3mm) specimens were employed with a span of 24mm.

The machined notch was 2mm deep, following some initial trials, to compare machined and razor sharpened notches, the latter approach was used for all the remaining tests. In a welded specimen, the notch was located in the centre of the weld zone, with the crack running in the through-thickness direction of the welded plate. Tests were conducted at room temperature (20–23°C) in three point bending at a loading rate of 10mm/min, using the test method described in the draft ISO standard for LEFM test on plastics.

Fracture mechanics tests were carried out in PEEK parent material and welds, unaged, aged in air at 250°C for 14 days and in boiling water for 14 days.

4.4.3. Measurement of residual stresses

4.4.3.1. Specimens

Residual stress measurements were made or attempted in fifteen specimens. These are listed in Table 2. The parent materials for the specimens were Victrex 450G PEEK, Victrex 4100G PES, Terluran 967K ABS, Ultem 1000 PEI and Valox 420 SEO glass reinforced polyester.

One of the PEEK specimens was a plain piece of parent material. All the other specimens were composed of two approximately square pieces of material welded together to form a rectangular panel with a central weld. The welded specimens were 140 to 167mm wide (across the weld), 65 to 85mm long (along the weld) and 3.0 to 3.4mm thick. The small variations in specimen dimensions are not expected to have had a significant effect on the residual stresses due to the weld.

Measurements of residual stresses in four of the PEEK specimens were made with the specimens in the as-welded condition, and between 5 and 28 months after welding. Measurements in a further four PEEK specimens were made after ageing in boiling water for 8 and 72 hours, and in air at 250°C for 8 and 72 hours. The measurements in the PEI and GRP specimens were made in the as-welded condition within 8 hours of welding. The measurements on the PES and ABS specimens were made in the as-welded condition, more than four years after welding.

4.4.3.2. Measurement procedure

The standard measurement procedure which was applied to 12 of the specimens, as listed in Table 2, was as follows:

1. The weld flash was removed down to 0.05mm from the surface using a single-fluted end mill.

2. The remaining 0.05mm of weld flash was removed by hand using 320 to 600 grit size 'wet and dry' abrasive papers.

3. The surfaces were cleaned prior to strain gauge attachment.

4. Two uniaxial type GFLA-3-70 strain gauges were bonded to the specimen using Kyowa type CC33A adhesive. The gauge layout is shown in Fig.5. A dummy gauge was attached to an additional specimen.

5. The gauges were connected to a Kyowa type UCAM-10A 10 channel pulsed excitation data logger. The specimens and instrumentation were moved to a temperature controlled room.

6. A 20mm wide strip containing the strain gauges was sawn out from the specimen using a junior hacksaw. The strain changes due to the removal of the strip were recorded.

7. The strain gauges were removed from the central strip. A new gauge was attached to the edge of the strip, as shown in Fig.6, and connected to the data logger.

8. Eight or nine layers of material of depth 2mm were machined from the edge opposite to the gauged edge using a single fluted end mill. The accumulated strain and the depth of the specimen were recorded after each layer removal.

9. The distribution across the weld of the longitudinal residual stresses (parallel to the welding direction) was calculated from the measured strain changes using TWI program BRSL-PC.

Program BRSL-PC is a standard TWI program for calculation of residual stresses from sectioning measurements. It is designed for measurement of the through-thickness variation of residual stresses, but can be used, as in the present instance, to give the variation across the width of the removed strip. The residual stresses in the original intact specimen are calculated as the sum of the stresses relaxed during the removal of the strip (stage 6) and the stresses relaxed during the removal of layers from the edge of the strip (stage 8). The former is calculated directly from the relaxed strains during strip removal, averaged from the gauges on the front and back faces. The stresses relaxed during layer removal are calculated on the basis of the assumption that the strip behaves like a beam during layer removal, with plane sections remaining plane. The program calculates the average stress in each layer removed, making due compensation for the stress changes due to the removal of previous layers.

The 'trial' procedure applied to the specimen containing weld 7199H75 was similar to the standard procedure, except that three 10mm wide strips were removed (Fig.7). The two outer strips were not layered. The central strip, including the weld, was layered with 1mm deep layers.

Since the weld flash was removed from all the specimens subjected to the trial or standard procedures, the measured stresses represent the residual stresses in the specimens after removal of the weld flash. A special procedure applied to two of the ABS specimens, was designed to investigate the effect of flash removal. The procedure was as follows:

1. The flash was removed from weld 7044H104, but left intact on weld 7044H105.

2. Four strain gauges were applied to each specimen (Fig.8).

3. 20mm wide strips containing the strain gauges were sawn out from the specimens, and the strain changes were recorded.

4. The removed strips were weighed.

5. The flash was removed from weld 7044H105. The strain change due to flash removal was recorded.

6. The strip containing weld 7044H105 was reweighed.

5. RESULTS AND DISCUSSION

5.1. Development of Welding Conditions for PEEK

5.1.1. Parameter investigation

The details of the welding parameters investigated and the resulting tensile strengths are given in Table 3.

The first series of welds investigated the effect of plate temperature. At all temperatures, the material in contact with the plate turned a dark brown colour and a small amount of molten material remained on the plate after welding. Since the lower temperature, 350°C, did not reduce either of these effects it was decided to carry out the remainder of the trials using a plate temperature of 400°C.

The second series of welds attempted to reduce the heating time to a minimum. However, the resulting welds were found to be weak and inconsistent. After testing, the fracture surface was shiny and smooth indicating that a cold weld had been made. The heating times were, therefore, increased considerably, up to a maximum of 120 seconds.

The final series of welds considered both heating time and heating pressure. The results show that good welds were achieved at all weld times over 60 seconds, and that the strongest joints were achieved at the higher heating and welding pressure of 0.3 N/mm^2.

From these trials, it can be concluded that a good welding condition for making hot plate welds in injection moulded PEEK plaques is as follows:

Heating Pressure	0.3 N/mm^2
Heating Time	90 seconds
Plate Temperature	400°C
Welding Pressure	0.3 N/mm^2
Cooling Time	30 seconds

5.1.2. Observations

Two important observations were made during the welding trials. Firstly, as the PEEK melted it turned amorphous, indicated by the dark brown appearance. As the two components were forged together, the PEEK crystallised slowly, with the final weld the same colour as the original parent material i.e. there were no visual signs of degradation of the PEEK at the weld. Secondly, a small amount of PEEK remained adhered to the hot plate after each weld. This was expected since there was no non-stick coating on the plate. However, if the process were to be used on this material as a production method, it would be necessary to develop a more practical solution to the problem than cleaning the plate between each weld.

5.1.3. Testing procedure

The original test specimen design is shown in Fig.3(a). However, for weld number 7199/H/1 this was found to be inadequate since the welded component failed away from the joint line. For this reason, the test geometry was slightly modified to that shown in Fig.3(b). Using this geometry, all welds failed at the joint line, allowing a better analysis of the results with respect to the effect of welding parameters. The test geometry used for each weld is detailed in Table 3.

5.1.4. Welds for degradation studies and residual stress measurements

Fifty welds were made at the best conditions identified in the parameter investigation. (See Section 5.1.1). The specimen design used is shown in Fig.3(b).

5.2. Ageing Studies

5.2.1. Tensile property measurements after air oven ageing

The results of tensile tests on unaged PEEK parent material and welds are given in Table 4. In all tests, the stresses rose to a maximum value and then decreased before ultimate failure. In polymer science, the first stress maximum is known as the yield strength σ_y, and the stress at breaking point is known as the tensile strength at break σ_b.

In most cases in the present project, the tensile strength at break was lower than the yield strength, but in a few cases the stresses increased at high strain, and the tensile strength at break was equal to or higher than the yield strength.

The effects of ageing at 250°C in air on the tensile properties of PEEK parent materials and welds are shown in Tables 5 and 6 and Figs. 9 – 11. Yield (σ_y), tensile strength at break (σ_b) and elongation at break of these materials were determined at different ageing times. For parent (Table 5), yield strength increased by about 19%, σ_b by about 11%, whereas the elongation at break dropped from 6.6mm to 1.8mm after 140 days ageing in air at 250°C. These changes suggest that PEEK stiffens during the ageing process resulting in a drop in ductility. The mechanism of degradation probably involved branching leading to crosslinking and loss in flexibility.

As shown in Table 6, both yield and tensile strength for hot plate welds of PEEK increased when ageing was increased from 14 to 140 days. There was 18% increase in σ_y and 27% increase in σ_b with ageing. The elongation at break decreased from 1.8mm to 0.6mm after the same period of ageing.

The most significant changes in tensile properties occurred within the first two weeks of ageing. The elongation at break of welds decreased by 68% and parent material by 29% after two weeks ageing. In general, this increase in σ_b and drop in elongation during ageing was more significant for welds than the parent materials, indicating greater extent of degradation in welds.

5.2.2. Tensile Property Measurements after Water Immersion

The effects of immersion in boiling water on tensile properties of PEEK parent materials and welds are given in Tables 7 and 8, and Figs. 12 – 14.

For parent material (Table 7), there was also very little change in σ_y, σ_b and elongation at break after 21 days ageing. Further exposure for 28 days did not cause any noticeable variation in tensile properties. The retention of σ_y and σ_b were 101% for both, whereas elongation at break dropped only by 8% after 21 days of ageing.

In the case of PEEK welds (Table 8) there was almost no change in σ_y, σ_b and elongation at break after 21 days ageing in boiling water.

5.2.3. Fracture Mechanic Tests

The results of fracture mechanic tests in terms of the fracture parameter J_0max are shown in Table 9 and Fig.15.

A photograph of a typical specimen after the fracture test is shown in Fig.4. It is apparent that the crack did not initiate at the notch tip on the weld centreline, but propagated parallel to the weld, at about 0.5mm from the centreline. It is particularly striking that the cracks initiated at the side of the notch and not at the sharpened notch tip. It is apparent that the weld was relatively strong and tough, but that there were a zone of weakness at about 0.5mm from the weld centreline.

None of the specimens exhibited true linear elastic behaviour at ambient temperature. While reducing the test temperature to say −40°C may result in linear elastic behaviour by increasing the yield stress and hence the constraint at the crack tip, it was felt that this was somewhat artificial as this material is likely to be used in service at ambient temperature or higher. Hence, it was decided to characterise the material by determining the fracture parameter, J_0max, at the point of maximum load on the load versus load–point displacement trace.

This parameter can be evaluated from the expression:

$$J_o\max = \frac{\eta U_{max}}{B(w-a_o)}$$

where:

U_{max}	=	area under load–point displacement diagram up to P_{max}
P_{max}	=	maximum load attained in the test
B	=	specimen thickness
W	=	specimen width
a_0	=	depth of razor sharpened notch
η	=	constant = 2 for SENB specimens.

The fracture mechanics test results are plotted in Fig.15. The first three sets of results were from the parent specimens as–received, after exposure in air at 250°C, and after exposure to boiling water (coded ARP, A250P and BWP) and the second three sets of results were for welded specimens in the same three conditions (coded ARW, A250W and BWW). For both parent and welded specimens, there was a small decrease after exposure in air at 250°C and a small increase in the average value of J_0max after exposure to boiling water. However it can be seen from the figure that the changes in toughness were not significant compared with the scatter in the results, particularly for the welded specimens.

5.3. Residual Stresses

5.3.1. Development of measurement procedure

The residual stresses measured by the trial procedure in a welded PEEK specimen, and by the standard procedure in welded and unwelded specimens, are shown in Fig.16

The trial measurements showed that the maximum tensile stresses occurred at the weld centreline. The tensile zone (the zone containing tensile residual stresses) was wider than the 10mm width of the central strip within which detailed measurements were made. For the standard procedure, the width of the central strip was increased to 20mm, to permit detailed measurements to be made over the whole width of the tensile zone.

It is interesting to compare the width of the zone containing tensile residual stresses with the width of the visible weld zone on microtome section (see Fig.17), which was found to be 0.1mm. Hence, it is apparent that the width of the tensile zone is at least two orders greater than the width of the visible weld zone.

The peak stresses measured by the standard procedure were less than those measured by the trial procedure, and the tensile zone was narrower. The measurements using the standard procedure were made 12 months after welding, while the trial procedure was applied five months after welding. It is assumed that the decrease in residual stresses was due to stress relaxation by creep in the older specimen. It would be desirable to make additional measurements using the standard procedure in specimens aged 0 and 5 months, in order to establish the initial stress state and to confirm the results for a 5 month old specimen.

The results for the unwelded specimen showed very low compressive stresses in the eight central measurement locations, in the range −0.5 to −1.4N/mm^2, and significant tensile stresses, up to 3.2N/mm^2, at the outer measurement positions corresponding to the centres of the first and last layers of the layer removal procedure. As there is no obvious reason why stress peaks should be present at these particular locations in the original specimen, the measured stresses probably represent residual stresses generated at the edges of the 20mm wide strip by the sawing operation, and do not represent the original stresses in the specimen.

Equally high stresses at the outer locations were obtained using the standard procedure in the welded specimen. These stresses are also presumably erroneous. For subsequent specimens, the operator was requested to saw out the central strip as gently as possible. None of the remaining specimens showed the sharp stress reversal obtained at the edges of the standard/welded specimen in Fig.16, and hence it appears that the error attributed to sawing stresses had been minimised.

Although sawing stresses at the edge of the strip must generate balancing stresses at the centre of the strip, the measurement procedure automatically compensates for this effect, and the sawing stresses do not cause any error at the central locations.

The compressive stresses observed at the central locations in the unwelded specimen may represent the original stresses in this region, or could be caused by experimental error associated with the layering process.

In summary, the experimental errors associated with the standard measurement procedure may be up to +3N/mm^2 at the two outer measurement locations, and between 0 and −1.5N/mm^2 at other locations. The errors associated with the measurement accuracy of the strain gauge instrumentation are negligible compared with these errors.

It was observed that there was a drift of up to ±7% of the final strain after layering was complete. This drift may have been associated with equalisation of temperatures in the remaining material, or time dependent relaxation of the remaining stresses. The observed drift corresponds to an uncertainty of up to ±0.8N/mm^2 in the stress measured by layering.

5.3.2. Effect of ageing of PEEK at ambient temperature

Figure 18 shows the measured residual stresses in four welded PEEK specimens after ageing in air at ambient temperature for between 5 and 28 months. The maximum measured stresses were as follows:

5 months	15.2N/mm^2
12 months	12.2N/mm^2
21 months	12.3N/mm^2
28 months	11.7N/mm^2

The maximum stresses and stress distributions for specimens aged 12, 21 and 28 months are very similar. It appears that there was no further stress relaxation after 12 months.

In engineering terms, time dependent relaxation of residual stresses can be regarded as a creep phenomenon. The application of stress in thermoplastic materials causes time–dependent plastic straining due to the rearrangement of the molecular chains. These plastic strains relieve the residual elastic strains and stresses, and hence the rate of relaxation decreases with time.

5.3.3. Effect of ageing of PEEK at elevated temperature

Figure 19 shows the measured residual stresses in four welded PEEK specimens after ageing in boiling water and at 250°C in air for 8 and 72 hours. The maximum residual stresses were as follows:

Boiling water x 8 hours	11.9N/mm^2
Boiling water x 72 hours	12.5N/mm^2
250°C x 8 hours	8.7N/mm^2
250°C x 72 hours	0.2N/mm^2

The elevated temperature treatments were applied 21 months after the weld had been made. The maximum stresses in the specimens aged in boiling water were very similar to those in the untreated 21 month old specimen (12.3N/mm^2). Hence, it appears that ageing in boiling water has no effect on the residual stresses.

However, there was a significant reduction in stresses after ageing at 250°C, down to 8.7N/mm^2 after 8 hours and down to effectively zero stresses after 72 hours. The low compressive stresses in this specimen may represent inherent experimental error, as discussed in Section 4.3.1.

Two of the stress distributions show small fluctuations. It is considered likely that these fluctuations are caused by experimental error, such as thermal strains during layer removal, rather than by genuine discontinuities in the stress fields.

5.3.4. ABS, PEI, GRP and PEEK

Figure 20 shows the residual stresses measured in the ABS, PEI and GR polyester specimens. The PES specimen cracked spontaneously when the central strip was sawn out, and no results were obtained. A second PES specimen was acquired, and this also was found to have cracked during sawing.

The peak stresses in the four materials for which results were obtained were as follows:

GR polyester	(0 months)	23.0N/mm^2
PEI	(0 months)	15.6N/mm^2
PEEK	(5 months)	15.2N/mm^2
ABS	(55 months)	6.2N/mm^2

The width of the zone of tensile stresses was about 12mm in the PEEK and GR polyester specimens, and about 8mm in the PEI and ABS specimens.

In Table 10, the maximum measured residual stresses are expressed as a proportion of the tensile strength σ_T and derived material parameters σ_{go} and σ_{so}. All the material properties in Table 10 are taken from manufacturers' data. The derived paraments σ_{go} and σ_{so} represent the stress corresponding to fully restrained uniaxial shrinkage from the glass transition temperature T_g and the softening temperature T_s to room temperature T_o, calculated as follows:

$$\sigma_{so} = E\alpha\ (T_s - T_o)$$

$$\sigma_{go} = E\alpha\ (T_g - T_o)$$

It can be seen that the maximum measured stresses were between 12 and 20% of the tensile strength in each material. This behaviour is very different from that at welds in steels and most other metals, where the maximum longitudinal residual stress is typically approximately equal to the yield strength, which is usually between 60 and 90% of the tensile strength. Yield magnitude residual stresses occur in welds because the shrinkage strain between the 'softening temperature' (ie. the temperature above which the strength is very low) and room temperature is many times greater than the yield strain, and hence the weld metal and adjacent heated parent metal yields during cooling after welding. In the polymers investigated in this project, the shrinkage strains during cooling from T_g or T_s were less than the strains corresponding to the yield or tensile strengths, and hence the maximum residual stresses were less than the yield or tensile strength.

In the final column of Table 10, it can be see that σ_R/σ_{go} is highly variable and hence it appears that the shrinkage from T_g to T_o does not provide a good guide to the maximum residual stresses. In the penultimate column of Table 10, it can be seen that the maximum residual stresses were about 50% of σ_{so} in the PEEK and PEI and GR polyester specimens, and about 32% of σ_{so} in ABS. It is apparent that maximum residual stress are much better correlated with σ_{so} than with σ_{go}, although it should be noted that they were only equal to about half or less of the shrinkage stresses which would be accumulated during shrinkage from T_s to T_o. It may be that the size of the specimens was not great enough to provide full restraint against shrinkage parallel to the weld. Based on experience in metals, the minimum specimen length to achieve maximum residual stresses is about 12 times the width of the tensile zone, ie. about 180mm, compared with the present samples which were about 75mm long (along the weld).

Further investigation of the development of residual stresses during cooling after welding of polymers would be desirable. Measurement of the accumulation of stress in restrained uniaxial specimens during simulated welding thermal cycles would contribute to an understanding of the generation of residual stresses during welding.

5.3.5. Stresses in weld flash in ABS

The strain changes and weight data for the removal of the 20mm wide strips from the ABS specimens containing welds 7044H105 and 7044H104 are summarised in Fig.21. The mean stresses $\bar{\sigma}_s$ in the 20mm wide strips and $\bar{\sigma}_f$ in the weld flash were calculated as follows, taking E = 2400N/mm^2. From measured strain changes:

$$E\Delta\varepsilon_1 = \bar{\sigma}_{s2} - \bar{\sigma}_{s1} = \bar{\sigma}_{f2} - \bar{\sigma}_{f1} \qquad [1]$$

$$E\Delta\varepsilon_2 = \bar{\sigma}_{s3} - \bar{\sigma}_{s2} \qquad [2]$$

$$E\Delta\varepsilon_3 = \bar{\sigma}_{s5} - \bar{\sigma}_{s4} \qquad [3]$$

In removed 20mm strips without weld flash:

$$\bar{\sigma}_{s3} = \bar{\sigma}_{s5} = 0 \qquad [4]$$

For equilibrium, and assuming cross-sectional area is proportional to weight:

$$\bar{\sigma}_{s2}.W_s + \bar{\sigma}_{f2}.W_f = 0 \qquad [5]$$

Then, from Eq.[3] and [4], the mean stress $\bar{\sigma}_{s4}$ in the 20mm wide strip containing weld 7044H104, before removal from the plate, is obtained as:

$$\bar{\sigma}_{s4} = -E\Delta\varepsilon_3 - 2400 \times -389 \times 10^{-6} = 0.9 N/mm^2$$

From Eq.[2] and [4], the mean stress $\bar{\sigma}_{s2}$ in the 20mm wide strip containing weld 7044H105, after removal from the plate but before removal of the flash, is obtained as:

$$\bar{\sigma}_{s2} = -E\Delta\varepsilon_2 = -2400 \times 284 \times 10^{-6} = -0.7 N/mm^2$$

From Eq.[5], the mean stress $\bar{\sigma}_{f2}$ in the flash of weld 7044H105, after removal of the strip from the plate, is obtained as:

$$\bar{\sigma}_{f2} = -\bar{\sigma}_{s2} W_s / W_f = 0.7 \times 4.26/0.63 = 4.6 N/mm^2$$

From Eq.[1], the mean stresses $\bar{\sigma}_{s1}$ in the 20mm wide strip and $\bar{\sigma}_{f1}$ in the flash of weld 7044H105, before removal from the plate, are obtained as:

$$\bar{\sigma}_{s1} = \bar{\sigma}_{s2} - E\Delta\varepsilon_1 = -0.7 - 2400 \times (-454 \times 10^{-6}) = 0.4 N/mm^2$$

$$\bar{\sigma}_{f1} = \bar{\sigma}_{f2} - E\Delta\varepsilon_1 = 4.6 - 2400 \times (-454 \times 10^{-6}) = 5.7 N/mm^2$$

Hence the mean longitudinal residual stress in the flash of weld 7044H105 is estimated to have been 5.7N/mm^2.

Comparison of $\bar{\sigma}_{s1}$ and $\bar{\sigma}_{s4}$ shows that the mean stresses in the 20mm wide central strip were 0.5N/mm^2 lower at weld 7044H105 with the weld flash intact than at weld 7044H104 with the flash removed. Hence it is predicted that the residual stresses in a specimen with an intact weld flash would have been 0.5N/mm^2 lower than those measured using the standard procedure at weld 7044H106 after removal of the weld flash. The peak stress measured at weld 7044H106 was 6.2N/mm^2, and is predicted to have been 6.2 – 0.5 = 5.7N/mm^2 before removal of the flash. This is exactly equal to the estimated mean stress in the flash in weld 7044H105. Hence it is concluded that the stress in the flash is equal to the peak stress in the plate. It is further concluded that stresses in the plate at welds with the flash removed are marginally greater than those with the flash intact. If this conclusion is also valid for other plastic materials, then the stresses measured in this project in specimens with the flash removed are marginally greater than those at welds with the flash intact.

6. CONCLUSIONS

1. The main part of this work was based on development of procedures and measurements of residual stresses in welds with the following main conclusions:

   (i) a procedure for the measurement of the variation of longitudinal residual stresses across the weld in plastic specimens has been developed. The estimated experimental error associated with the procedure is between 0 and $-1.5N/mm^2$ at all locations except the outer edges of the 20 mm wide central strip;

   (ii) the maximum measured longitudinal residual stress at welds in PEEK was $15N/mm^2$ in a specimen aged in air at ambient temperature for five months after welding; the maximum stress was reduced to $12N/mm^2$ in a PEEK specimen aged for 12 months, but showed no further reductions in specimens aged for 21 and 28 months;

   (iii) there was no further reduction of residual stresses in PEEK specimens aged for up to 72 hours in boiling water; but the maximum residual stresses in PEEK specimens aged in air at 250°C reduced to $9N/mm^2$ after 8 hours and $0.2N/mm^2$ after 72 hours;

   (iv) the maximum measured residual stresses in GR polyester and PEI specimens immediately after welding were 23 and $16N/mm^2$ respectively; the maximum residual stress measured in an ABS specimens 55 months after welding was $6N/mm^2$; the maximum measured residual stresses in each material were between 12 and 20% of the tensile strength.

   (v) the width of the zone of tensile residual stresses in PEEK was about 12 to 14mm, compared with the width of the visible weld zone, which was about 0.1mm.

2. The conclusions from ageing studies were as follows:

   (i) Ageing of tensile specimens in air at 250°C resulted in more significant deterioration in tensile properties of PEEK welds than parent materials. The tensile and yield strength generally increased in both cases whereas there was a reduction in elongation at break.

   (ii) Ageing in boiling water for 21 days did not cause significant changes in tensile properties of PEEK welds or parent materials.

   (iii) Fracture mechanic tests on single edge notched bend (SENB) specimens suggested there were small changes in fracture toughness of parent and welded PEEK specimens after exposure in air at 250°C and exposure to boiling water: however, these changes were not significant compared with the scatter in the results.

3. A good condition was established for hot plate welding of PEEK as follows:

heating pressure – 0.3N/mm^2, heating time – 90 seconds, plate temperature – 400°C, welding pressure – 0.3N/mm^2, cooling time – 30 seconds.

7. FURTHER WORK

A full description of the heat flow, plastic straining and generation of residual stresses during welding of polymers could be obtained by numerical modelling of the welding process using the finite element method. The residual stress data generated in this project could be used for experimental validation of numerical modelling.

8. ACKNOWLEDGEMENTS

The work was carried out within the core research programme of TWI which is jointly funded by the Information and Manufacturing Technology division of the Department of Trade and Industry, and the Industrial Members of TWI. Thanks are also due to colleagues at TWI, particularly Dr G E Hale for fracture mechanic tests, Miss N S Taylor for welding of PEEK, Mr B Kersey and Mr B A Martin for experimental work on ageing and stress measurements.

9. REFERENCES

1 Scott G, 'Atmospheric oxidation and antioxidants', Elsevier, New York and London, 1965.

2 Grassie N and Scott G, 'Polymer degradation and stabilisation', Cambridge University Press, 1985.

3 Ranby B and Rabek J F, 'Photo-degradation and photo-stabilisation of polymers', Wiley and Sons, 1975.

4 Tavakoli S M, 'Polymer degradation – fast or slow, how and why', TWI Bulletin 6, 417/6/91, November/December 1991, 136–139.

5 Tavakoli S M, 'The effect of ageing on structure and properties of hot plate welds in acrylonitrile butadiene styrene', TWI members report 440/1992, January 1992.

6 Tavakoli S M, 'The effect of ageing on structure and properties of hot plate welds in acrylonitrile butadiene styrene', paper presented at ANTEC 94, 1–5 May, 1994, San Francisco, USA.

7 Tavakoli S M, 'The effect of ageing on tensile properties and structure of linear friction and hot plate welds in polyetherethersulphone', TWI members report 481/1993, December 1993.

8 Tavakoli S M, 'Effects of ageing on tensile properties and structure of linear friction and hot plate welds in polyetherethersulphone', paper presented at ANTEC 95, 'The plastics challenge – a revolution in Education', 7–11 May 1995, Boston, USA.

9 Haworth B, Hindle C S, Sandilands E J and White J R, 'Assessment of internal stresses in injection–moulded thermoplastics', Plastics and Rubber Processing and Applications 2, 59–71, 1982.

10 Pham H T, Bosnyak C P and Sehanobish K, 'Residual stresses in injection moulded polycarbonate rectangular bars', Proc. ANTEC '91, 1703–1705, 1991.

11 Manson J–A and Seferis J C, 'Internal stress determination by process simulated laminates', Proc. ANTEC '87, 1446–1449, 1987.

12 Jeronimidis G and Parkyn A T, 'Residual stresses in carbon fibre–thermoplastic matrix laminates', Journal of Composite Materials, 401–415, 22 May 1988.

13 Chapman T J, Gillespie J W, Pipes R B, Manson J–A and Seferis J C, 'Prediction of process–induced residual stresses in thermoplastic composites', Journal of Composite Materials, 616–643, 24 June 1990.

14 Sullivan T and Matsuoka S, 'Determination of thermally–induced stresses in plastics by photoelastic techniques', Proc. ANTEC '89, 790–795, 1989.

15 Young R J, Day R J, Zakikhani M and Robinson I M, 'Fibre deformation and residual thermal stresses in carbon fibre reinforced PEEK', Composites Science and Technology 34, 243–258, 1989.

16 Filiou C, Galiotis C and Batchelder D N, 'Residual stress distribution in carbon fibre/thermoplastic matrix pre–impregnated composite tapes', Composites 23, 28–38, 1 January 1992.

17 Tavakoli S M, 'Hot plate welding of PEI and a reinforced thermoplastic polyester', unpublished work.

18 Prime R B, 'Thermo–oxidative decomposition of poly (ether–ether–ketone),' Journal of Polymer Science: Part C, Polymer Letters, Vol. 24, 641–644, 1986.

19 Day M, Suprunchuk T, Cooney J D and Wiles D M, 'Thermal degradation of poly (aryl–ether–ether–ketone) (PEEK)' A Differential Scanning Calorimetry Study, Journal of Applied Polymer Science, Vol. 36, 1097–1106, 1988.

20 Hay J N and Kemmish D J, 'Thermal Decomposition of poly (aryl ether ketones)', Polymer, Vol.28, p.2047, Nov, 1987.

21 Day M, Cooney J D, Wiles D M, 'A kinetic study of the thermal decomposition of poly (aryl–ether–ether–ketone) PEEK in nitrogen', Polymer Engineering and Science, Vol. 29, No.1, 19, mid–January 1989.

22 Jonas A and Legras R, 'Thermal stability and crystallisation of poly (aryl ether ether ketone), Polymer, Vol 32, No.15,2691, 1991.

23 Sasuga T and Hagiwara M, 'Radiation deterioration of several aromatic polymers under oxidative conditions', Polymer, Vol.28, 1915, October 1987.

24 Sasuga T, Hayakawa N, Yoshida K and Hagiwara M, 'Degradation in tensile properties of aromatic polymers by electron beam irradiation', Polymer, Vol 26, 1039, July 1985.

25 Funk J G and Sylees G F, 'Space radiation effects on poly (aryl–ether–ketone) thin films and composites', SAMPE quarterly 19, April 1988.

26 Munro H S, 'Surface photo–oxidation of phenoxy resin and polyetheretherketone', Polymer Degradation and Stability, 19, 353–363, 1987.

27 Ma C C M and Yur S W, 'Environmental effect on the water absorption and mechanical properties of carbon fibre reinforced PPS and PEEK composites', ANTEC, 1389, 1496, 1989.

28 Benchman A and Isayer A I, 'Water absorption of some thermoplastic composites', Sampe Journal, Vol 27, No.4, July/August 1991 p30–36.

Table 1 Function of machine controls in hot plate welding

Control	Function
Hot Plate Temperature	Thermostatic control of heating element (50–450°C)
Heating Pressure	Controls axial pressure applied during heating (0–600 N)
Heating Time	Controls the duration of component contact with the hot plate (0–60 sec)
Displacement Limit Stops	Limits displacement during heating (0–1.0mm)
Weld Consolidation Pressure	Controls axial pressure applied during weld consolidation (0–600 N)
Weld Consolidation Time	Controls the duration of applied axial pressure through weld consolidation period (0–60 sec)

Table 2 Summary of residual stress measurement specimens

Weld No.	Material	Age (Months)	Condition/Treatment	Dimensions (mm)	Measurement Procedure	Note
7199H75	PEEK	5	As–welded	142 x 75 x 3.0	Trial	
99H55	PEI	0	As–welded	167 x 85 x 3.0	Standard	
V99H119	GR Polyester	0	As–welded	170 x 85 x 3.0	Standard	
7199H74	PEEK	12	As–welded	140 x 75 x 3.0	Standard	
N/A	PEEK	N/A	Un–welded	150 x 75 x 3.0	Standard	
7199H53	PEEK	28	As–welded	144 x 74 x 3.1	Standard	
7199H108	PEEK	21	As–welded	142 x 76 x 3.1	Standard	
7199H91	PEEK	21	99°C x 8 hours	142 x 74 x 3.1	Standard	Treatment in water
7199H112	PEEK	21	99°C x 72 hours	140 x 76 x 3.0	Standard	Treatment in water
7199H111	PEEK	21	250°C x 8 hours	142 x 77 x 3.1	Standard	Treatment in air
7199H110	PEEK	21	250°C x 72 hours	141 x 75 x 3.3	Standard	Treatment in air
7044H168	PES	50	As–welded	140 x 73 x 3.4	Standard	Cracked after sawing
7044H106	ABS	55	As–welded	155 x 65 x 3.1	Standard	
7044H105	ABS	55	As–welded	155 x 65 x 3.1	Special	Flash not removed
7044H104	ABS	55	As–welded	155 x 65 x 3.1	Special	

Table 3 Details of welding parameters and tensile strengths

Weld No	Pressure N/mm^2		Heat Time (sec)	Cool Time (sec)	Plate temp. (°C)	Displacement (mm)	Test Geometry	Tensile Strength (N/mm^2)	Comments
	Heating	Welding							
7199/H/1	0.2	0.2	25	30	400	2.5	a	81.3	Failed in jaws
7199/H/2	0.2	0.2	25	30	350	1.6	a	38.8	
7199/H/3	0.2	0.2	25	30	375	2.0	a	55.3	
7199/H/4	0.2	0.2	25	30	400	2.5	b	11.0	
7199/H/5	0.2	0.2	15	30	400	2.0	b	8.6	
7199/H/6	0.2	0.2	10	30	400	1.6	b		Failed during debeading
7199/H/7	0.2	0.2	5	30	400	1.2	b		Failed during debeading
7199/H/8	0.2	0.2	10	30	400	1.7	b	9.7	
7199/H/9	0.2	0.2	25	30	400	2.4	b	21.0	
7199/H/10	0.2	0.2	25	30	400	2.4	b	21.0	
7199/H/11	0.2	0.2	25	30	400	2.4	a	26.0	
7199/H/12	0.2	0.2	90	30	400	3.6	b	67.3	
7199/H/13	0.2	0.2	120	30	400	3.5	b	75.1	
7199/H/14	0.3	0.3	90	30	400	3.5	b	105.6	Some necking at joint
7199/H/15	0.3	0.3	120	30	400	4.1	b	106.5	Some necking at joint
7199/H/16	0.3	0.3	60	30	400	3.8	b	104.3	Some necking at joint
7199/H/17	0.2	0.2	60	30	400	3.6	b	90.1	

Table 4 Tensile properties of unaged PEEK parent material and hot plate weld

	Specimen Number	Yield Strength σ_y (N/mm^2)	Tensile Strength at Break (σ_b) (N/mm^2)	Elongation at Break (mm)
Parent Material	P02–25	105	87.7	6.4
	P02–26	105.6	90.4	7.4
	P02–27	106	90.3	6.0
Mean		105.5	89.5	6.6
Weld	T01–25	104.1	103.3	0.3
	T01–26	104.6	97.5	3.4
	T01–27	106.8	93.7	1.7
Mean		105.2	98.2	1.8

Table 5 Effect of ageing in hot air at 250°C on tensile properties of PEEK parent materials

Specimen Number	Ageing Time (Days)	Yield Strength (σ_y) (N/mm^2)	Retention of Original Value (%)	Tensile Strength at Break (σ_b) (N/mm^2)	Retention of Original Value (%)	Elongation at Break (mm)	Retention of Original Value (%)
P02–04	14	114.5	108	94.0	105	5.4	82
P02–05	14	115.9	110	96.9	108	4.2	64
P02–06	14	115.3	109	96.3	108	4.5	68
Mean		115.2	109	95.7	107	4.7	71
P02–10	28	114.6	109	97.9	109	3.5	53
P02–11	28	116.6	110	101.3	113	4.2	64
P02–12	28	116.3	110	99.7	111	4.0	61
Mean		115.9	110	99.6	111	3.9	59
P02–13	56	120	114	102	114	3.4	51
P02–14	56	119	113	102.3	114	3.6	54
P02–15	56	119	113	105	117	4.1	62
Mean		119.3	113	103.1	115	3.7	56
P02–19	112	119.3	113	100.8	112	1.8	27
P02–20	112	119.1	113	104.8	117	1.4	21
P02–21	112	121.2	115	105.9	118	2.4	36
Mean		119.9	114	103.8	116	1.9	28
P02–16	140	125.6	119	108.2	121	2.9	44
P02–17	140	126.2	120	114.8	128	1.9	29
P02–18	140	125.5	119	109.8	123	2.7	41
Mean		125.8	119	110.9	124	2.5	38

Table 6 Effect of ageing in hot air at 250°C on tensile properties of hot plate welds in PEEK materials

Specimen Number	Ageing Time (Days)	Yield Strength (σ_y) (N/mm^2)	Retention of original value (%)	Tensile Strength at Break (σ_b) (N/mm^2)	Retention of Original Value (%)	Elongation at Break (mm)	Retention of Original Value (%)
T01-04	14	115.8	110	112.8	115	0.6	33
T01-05	14	112.8	107	106.7	109	1.1	61
T01-06	14	115	109	114.9	117	0.5	28
Mean		114.5	109	111.5	114	0.7	41
T01-10	28	115.1	109	112.5	115	0.6	33
T01-11	28	116.3	110	112.2	114	1	55
T01-12	28	117.7	112	116	118	0.6	33
Mean		116.4	110	113.6	116	0.7	40
T01-13	56	118.9	113	118.1	120	0.8	44
T01-14	56	116.1	110	116.1	118	0.3	17
T01-15	56	117.9	112	117.9	120	0.1	5
Mean		117.6	111	117.4	119	0.4	22
T01-19	112	121.8	116	121.8	124	0.4	22
T01-20	112	119.5	114	119.5	122	0.3	16
T01-21	112	122.2	116	122.2	124	0.1	5
Mean		121.2	115	121.2	123	0.3	14
T01-16	140	122.4	116	122.4	125	0.5	28
T01-17	140	126.4	120	126.4	129	0.6	33
T01-18	140	126.1	119	126.1	128	0.8	44
Mean		125	118	125	127	0.6	35

Table 7 Effect of immersion in boiling water on tensile properties of PEEK parent materials

Specimen Number	Ageing Time (Days)	Yield Strength (σ_y) (N/mm^2)	Retention of Original Value (%)	Tensile Strength at Break (σ_b) (N/mm^2)	Retention of Original Value (%)	Elongation at Break (mm)	Retention of Original Value (%)
P02–01	7	106.6	101	89.1	99	4.02	61
P02–02	7	106.2	100	88.4	99	5.98	91
P02–03	7	106.5	101	87.8	98	6.67	101
Mean		106.4	101	88.4	99	5.5	84
P02–07	14	106.1	101	90.5	101	7.6	115
P02–08	14	106.6	101	89.5	100	6.7	101
P02–09	14	107.7	102	89.6	100	7.2	109
Mean		106.8	101	89.9	100	7.2	108
P02–22	21	106.7	101	87.4	98	6	91
P02–23	21	106.3	101	90.9	102	6	91
P02–24	21	107.7	102	91.2	102	6.2	94
Mean		106.9	101	89.8	101	6.1	92
P02–28	28	103.3	98	86	96	6	91
P02–29	28	101.9	97	87.7	98	6.5	98
P02–30	28	101.9	97	87.6	98	4.7	71
Mean		102.3	97	87.1	97	5.7	87

Table 8 Effect of immersion in boiling water on tensile properties of hot plate welds in PEEK

Specimen Number	Ageing Time (Days)	Yield Strength (σ_y) (N/mm^2)	Retention of Original Value (%)	Tensile Strength at Break (σ_b) (N/mm^2)	Retention of Original Value (%)	Elongation at Break (mm)	Retention of Original Value (%)
T01–01	7	107.4	102	94.3	96	2.5	139
T01–02	7	106.5	101	93.3	95	2.2	122
T01–03	7	107	102	95.9	98	1.6	89
Mean		106.9	102	94.5	96	2.1	117
T01–07	14	107	102	93.5	95	1.4	78
T01–08	14	107.4	102	94.6	96	1.8	100
T01–09	14	107.4	102	96.8	96	1.5	83
Mean		107.3	102	95	95.7	1.6	87
T01–22	21	107.6	102	96.2	98	1.6	88
T01–23	21	107.5	102	95.2	97	2.6	144
T01–24	21	109.4	104	104.4	106	1.4	78
Mean		108.2	103	98.6	100	1.9	103

Table 9 Fracture toughness data for PEEK welds

Specimen No.	Treatment	Condition	Fracture toughness J_{omax} (kj/m^2)
W01-07	As-received	Parent	19.45
W01-08	As-received	Parent	15.83
W01-09	As-received	Parent	16.92
W01-12	As-received	Parent	15.59
W01-13	As-received	Parent	16.15
W01-14	As-received	Parent	16.58
W01-15	As-received	Parent	20.30
W01-16	As-received	Parent	17.52
Mean ± δ			17.29 ± 1.72
W02-12	As-received	Weld	6.85
W02-13	As-received	Weld	12.12
W02-14	As-received	Weld	7.66
W02-17	As-received	Weld	6.62
W02-19	As-received	Weld	14.77
Mean ± δ			9.60 ± 3.6
WO1-04	14 days in air at 250°C	Parent	16.61
W01-05	14 days in air at 250°C	Parent	14.77
W01-06	14 days in air at 250°C	Parent	18.03
Mean ± δ			16.47 ± 1.63
W02-06	14 days in air at 250°C	Weld	3.32
W02-08	14 days in air at 250°C	Weld	15.03
Mean ± δ			9.17 ± 8.28
W01-01	14 days in boiling water	Parent	17.59
W01-02	14 days in boiling water	Parent	17.38
W01-03	14 days in boiling water	Parent	18.22
Mean ± δ			17.73 ± 0.44
W02-02	14 days in boiling water	Weld	10.53
W02-04	14 days in boiling water	Weld	9.19
Mean ± δ			9.86 ± 0.95

Table 10 Properties of parent materials (manufacturers' data) and measured maximum residual stresses

	Young's Modulus E, N/mm^2	Flexural Modulus N/mm^2	Tensile Strength, σ_T, N/mm^2 (1)	Coefficient of Expansion α, °C^{-1}	Vicat Softening Temperature T_s, °C	Glass Transition Temperature T_g, °C	Maximum Measured Residual Stress, σ_R, N/mm^2	σ_R/σ_T	σ_R/σ_{so} (5)	σ_R/σ_{go} (6)
PEEK Victrex 450G	3100	3660	92 (Y)	47 x 10^{-6}	200 (2)	143	15	0.16	0.58	0.88
PEI ULTEM 1000	3000	3300	90 (Y)	56 x 10^{-6}	219	215 (3)	16	0.18	0.48	0.50
GR Polyester Valox 420SE0	9000	7500	115 (B)	25 x 10^{-6}	213	45 – 50 (3)	23	0.20	0.53	3.83
PES Victrex 4100G		2600	84	55 x 10^{-6}	226	230 (4)	–	–	–	–
ABS Terluran 967K	2400		50	80 to 110 x 10^{-6}	107	105 (4)	6	0.12	0.32	0.32

Notes

(1) $\sigma_T(Y) = \sigma_y$ = stress at yield
$\sigma_T(B) = \sigma_b$ = stress at break
Condition not specified for PES and ABS

(2) No Vicat temperature available: value taken as temperature at which tensile strength was 10% of room temperature value.

(3) Verbal communication from manufacturer.

(4) "Typical value" quoted in Alger SM, "Polymer Science Dictionary", Elsevier Applied Science, 1989.

(5) $\sigma_{so} = E\alpha\,(T_s - T_o)$ with T_o = 23°C

(6) $\sigma_{go} = E\alpha\,(T_g - T_o)$ with T_o = 23°C

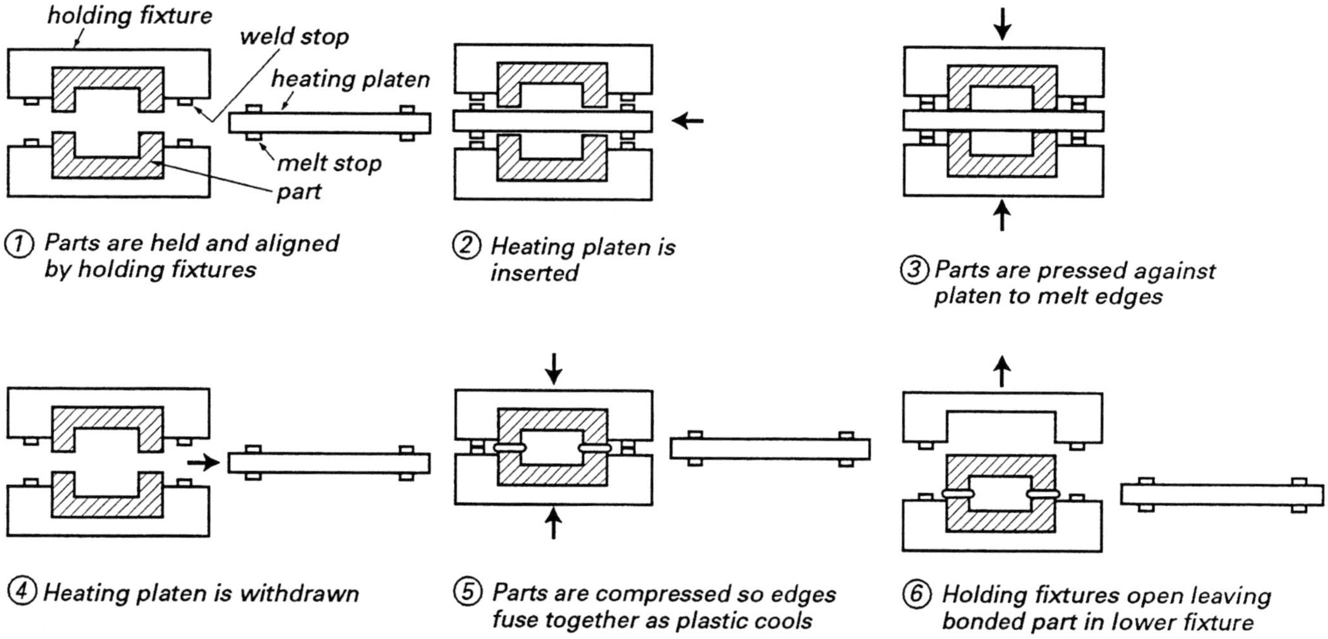

Fig. 1. Sequence of operations for hot plate welding of a pair of channel section parts.

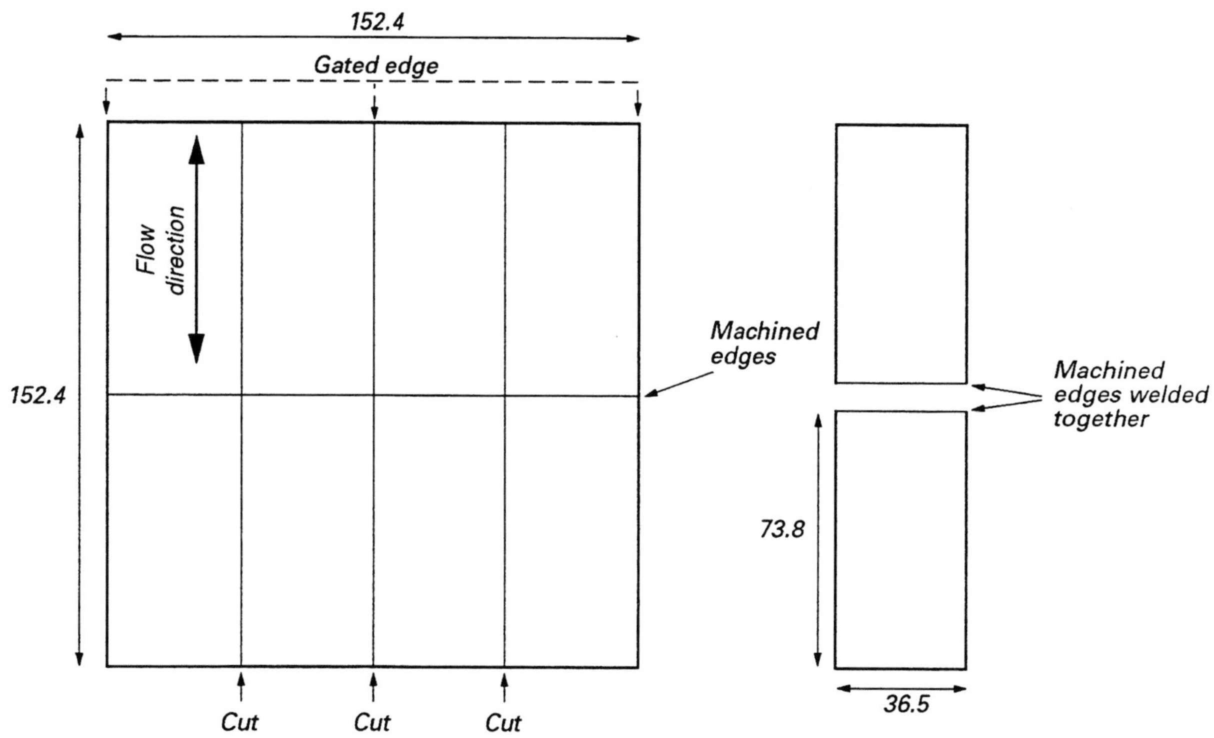

Fig. 2 (a). *Cutting diagram and specimen geometry for parameter investigation, units in mm.*

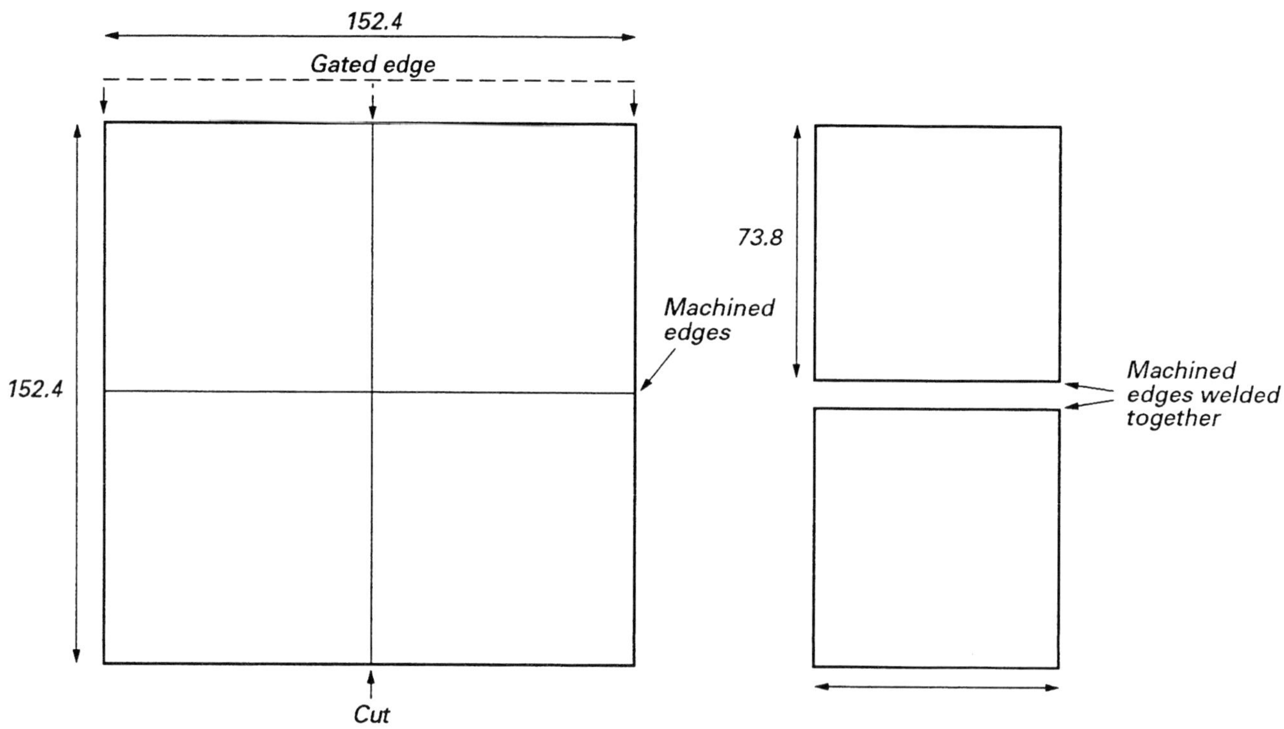

Fig. 2 (b). *Cutting diagram and specimen geometry for degradation study of welds, units in mm.*

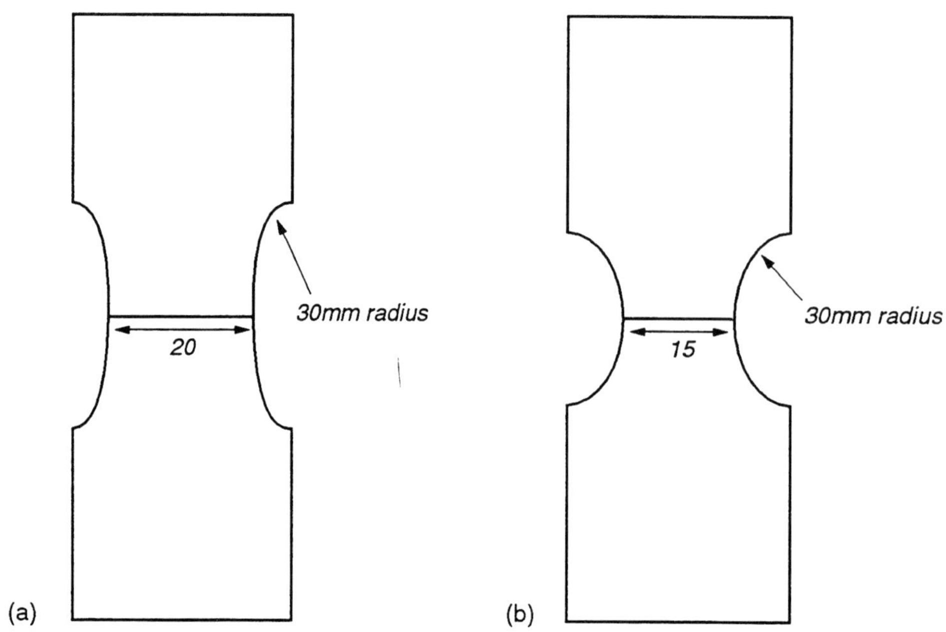

Fig. 3. Test piece geometry
(a) *Initial.*
(b) *Final.*

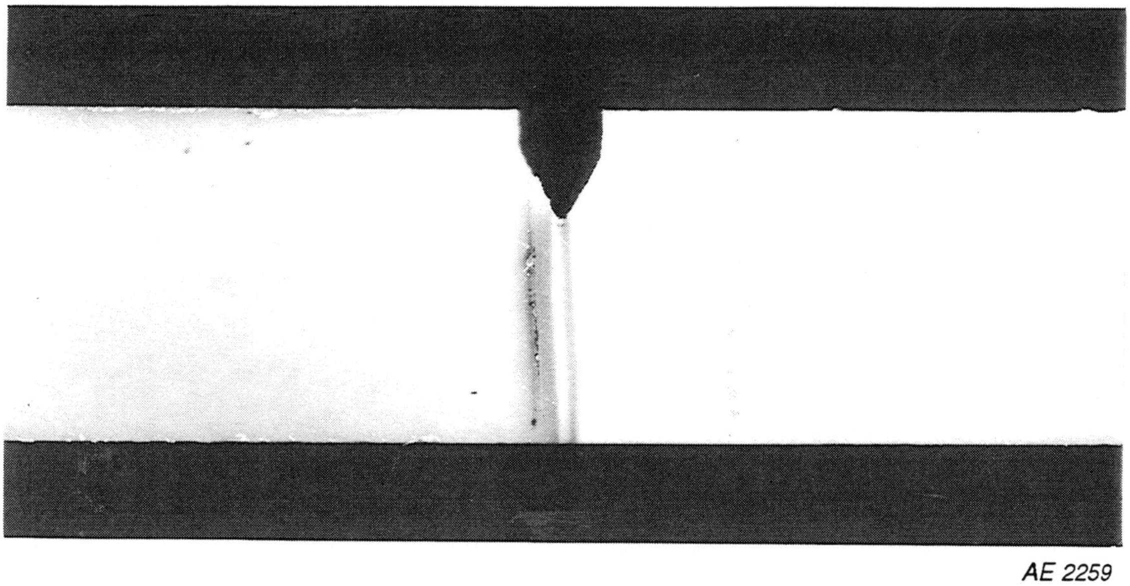

Fig. 4. A typical single edge notched bend (SENB) specimen of a hot plate weld in PEEK. Magnification x15.

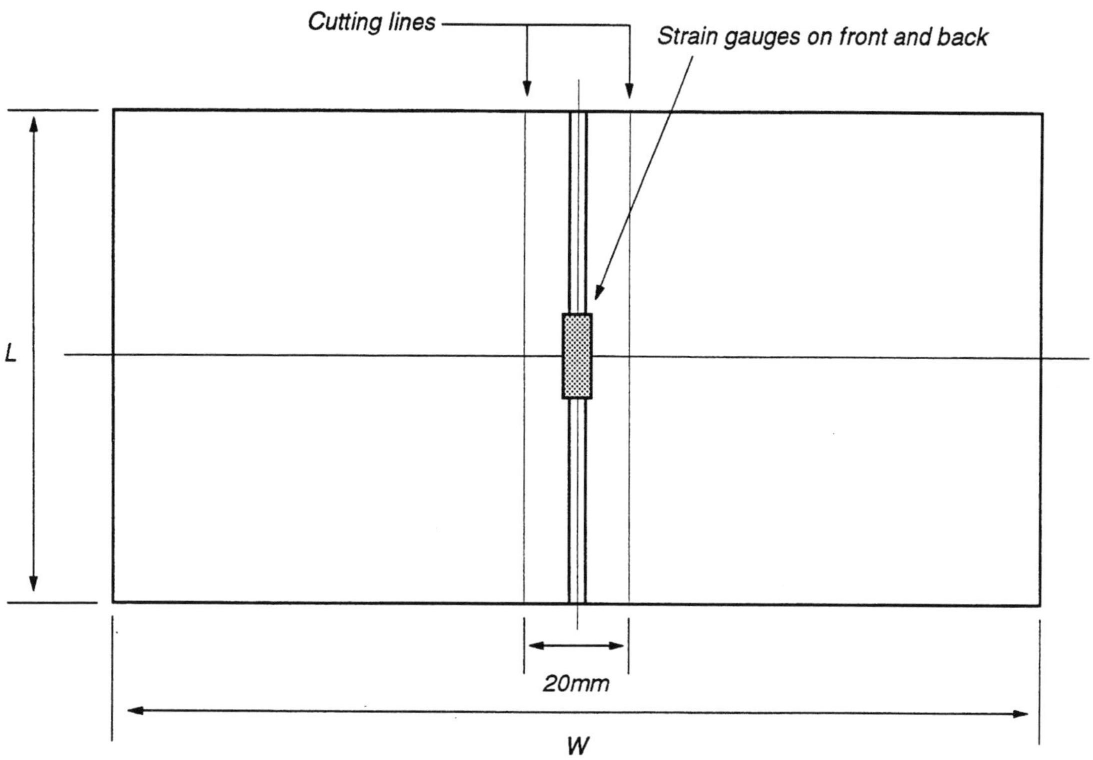

Fig. 5. Gauge locations and cutting lines for strip removal for standard residual stress measurement procedure.

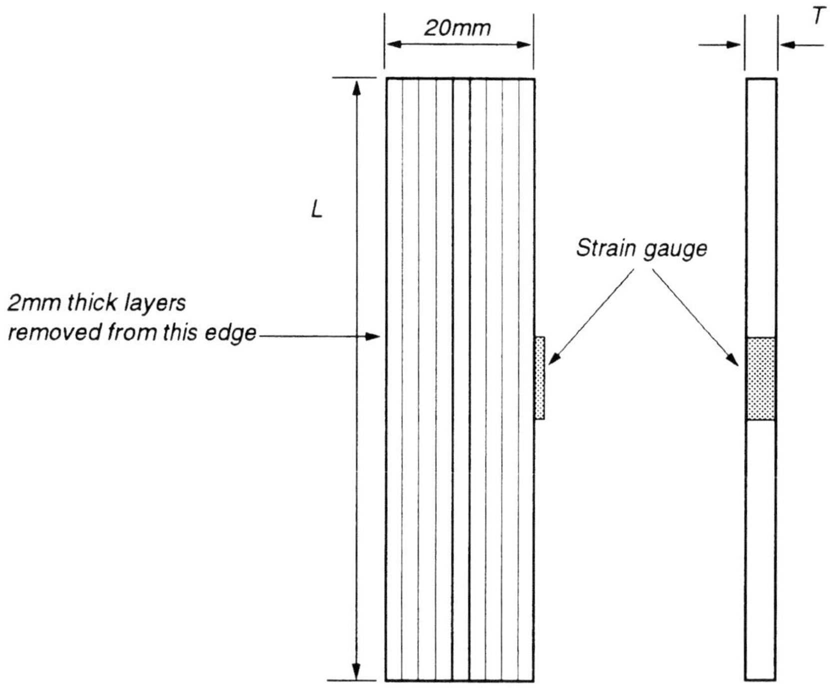

Fig. 6. Gauge location for layer removal for standard residual stress measurement procedure.

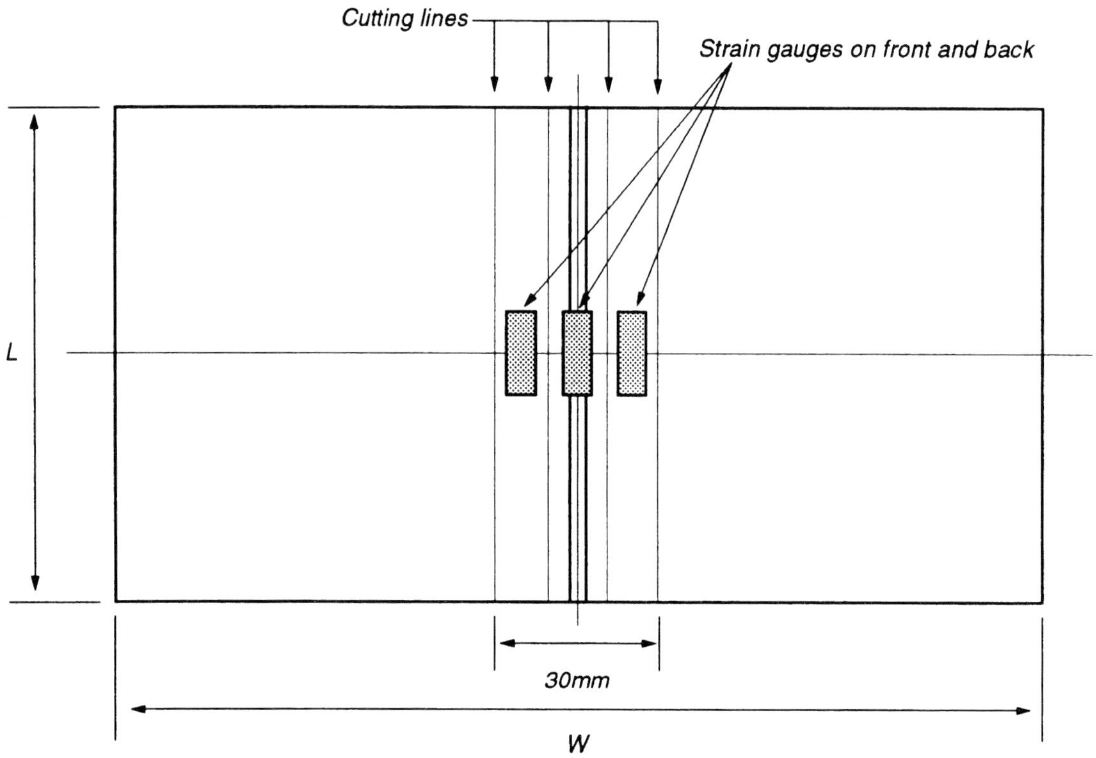

Fig. 7. Gauge locations and cutting lines for strip removal for trial residual stress measurement procedure applied to weld 7199H75.

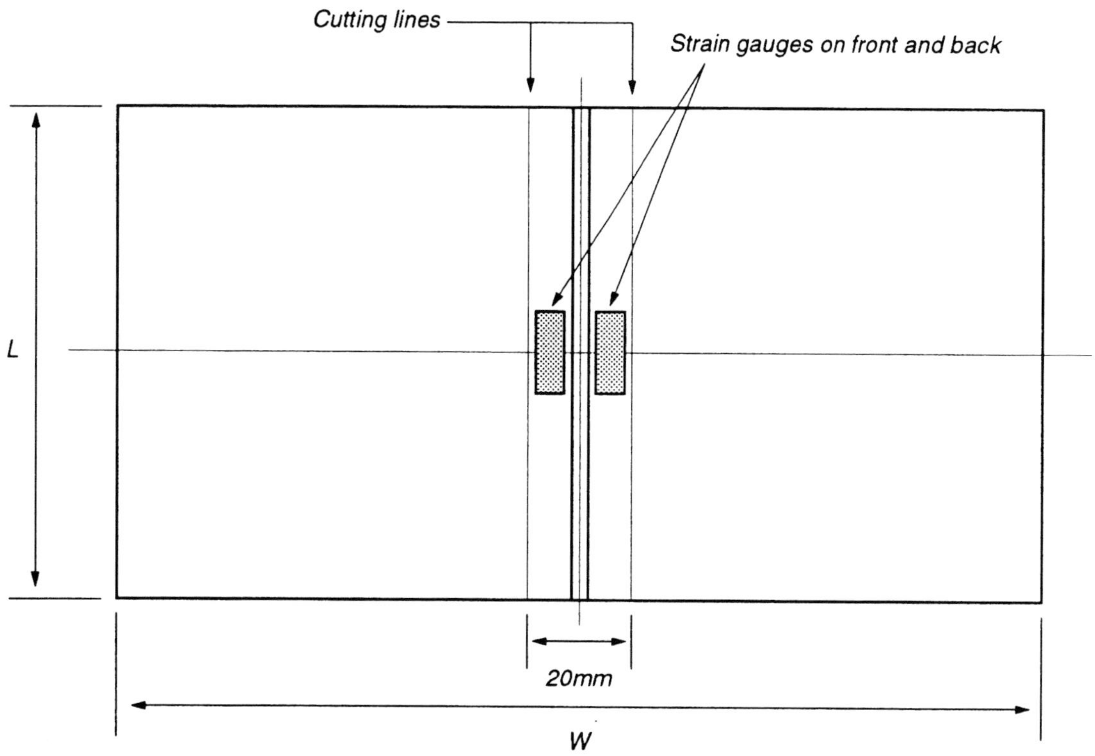

Fig. 8. Gauge locations and cutting lines for strip removal for special residual stress measurement procedure applied to ABS welds 7044H105 and 7044H104.

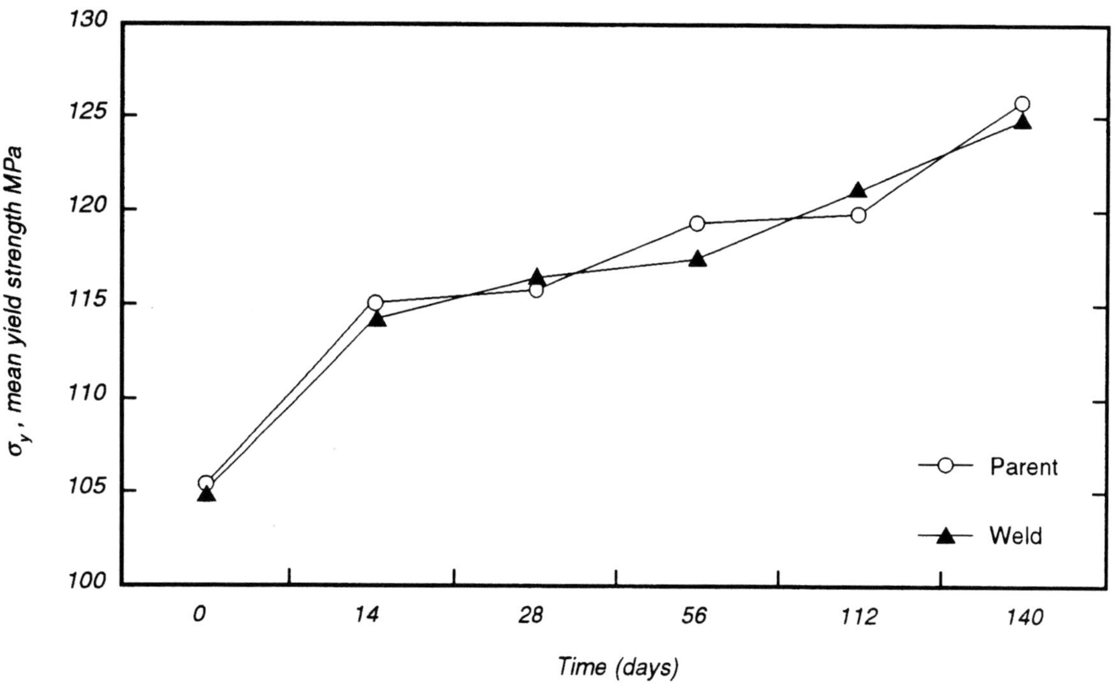

Fig. 9. Effect of aging in air at 250°C on yield strength of PEEK parent materials and hot plate welds.

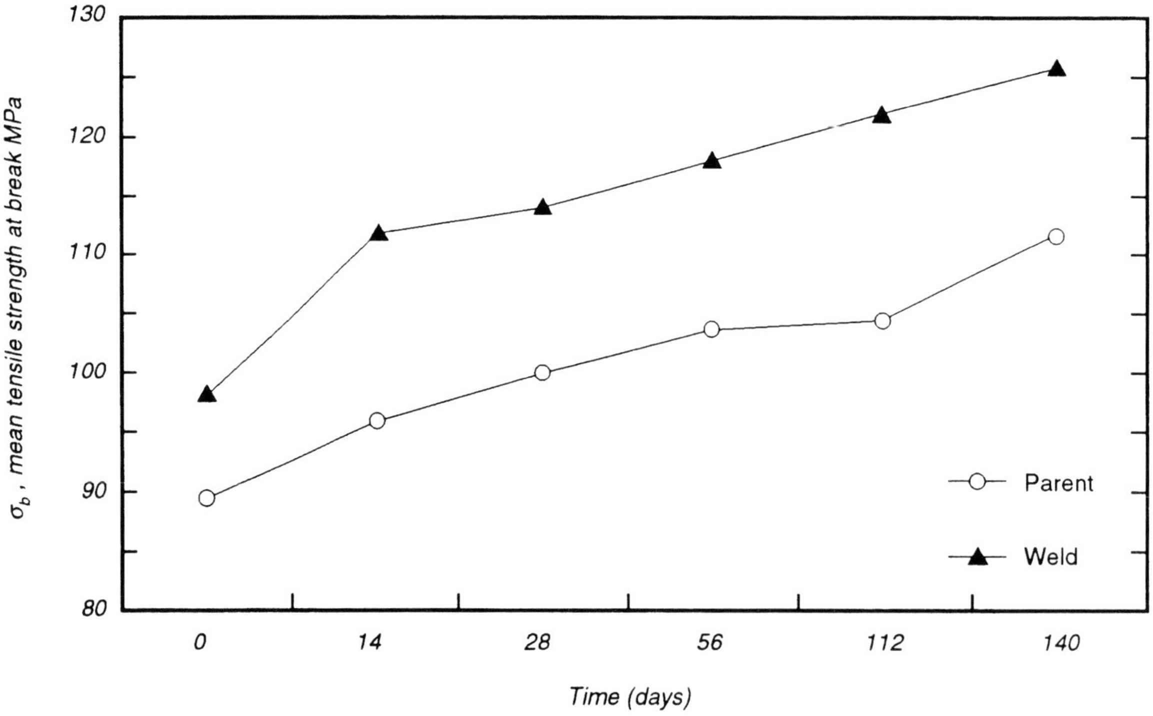

Fig. 10. Effect of aging in air at 250°C on tensile strength of PEEK parent material and hot plate welds.

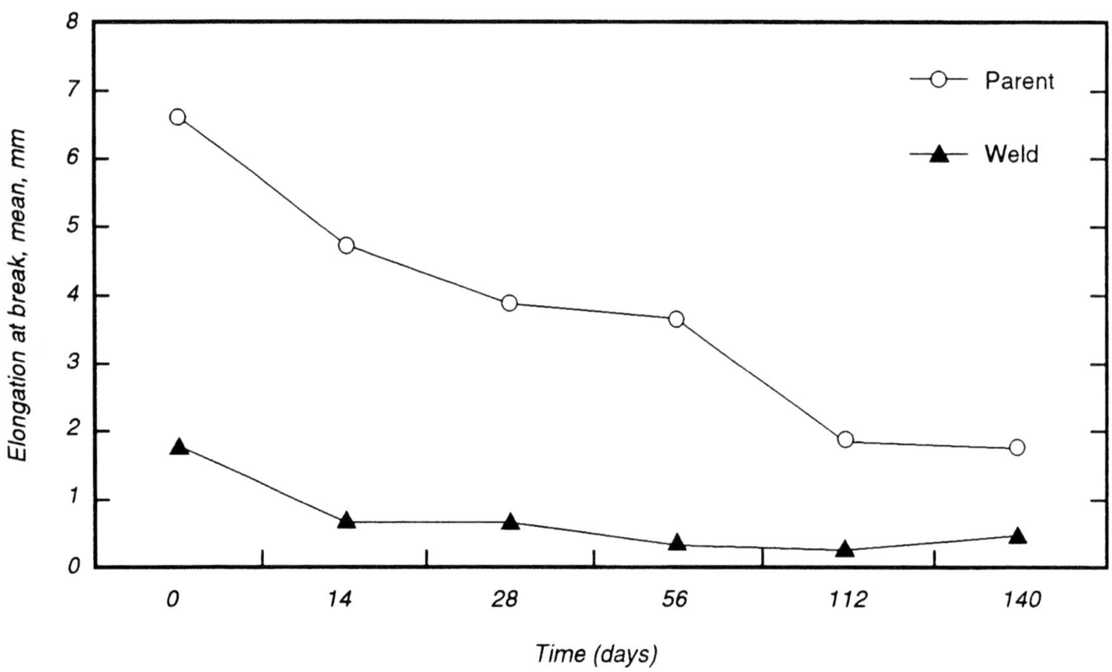

Fig. 11. Effect of aging in air on elongation at break of PEEK parent materials and welds.

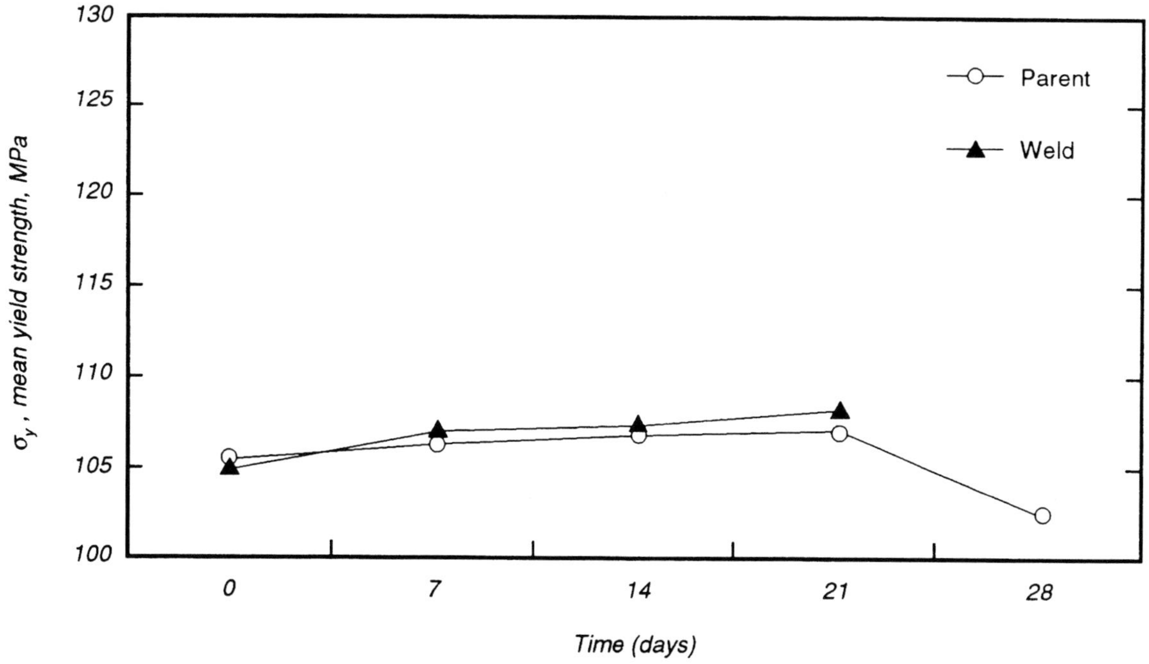

Fig. 12. Effect of immersion in boiling water on yield strength of PEEK parent material and hot plate welds.

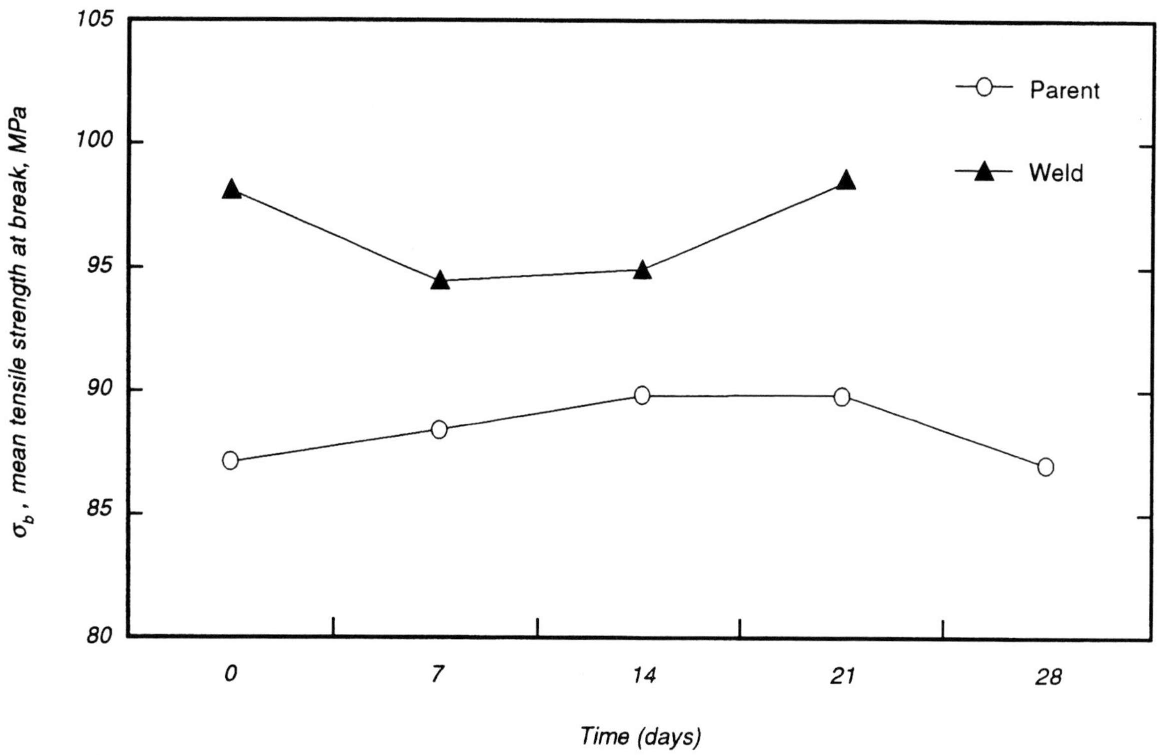

Fig. 13. Effect of immersion in boiling water on tensile strength at break of PEEK parent material and hot plate welds.

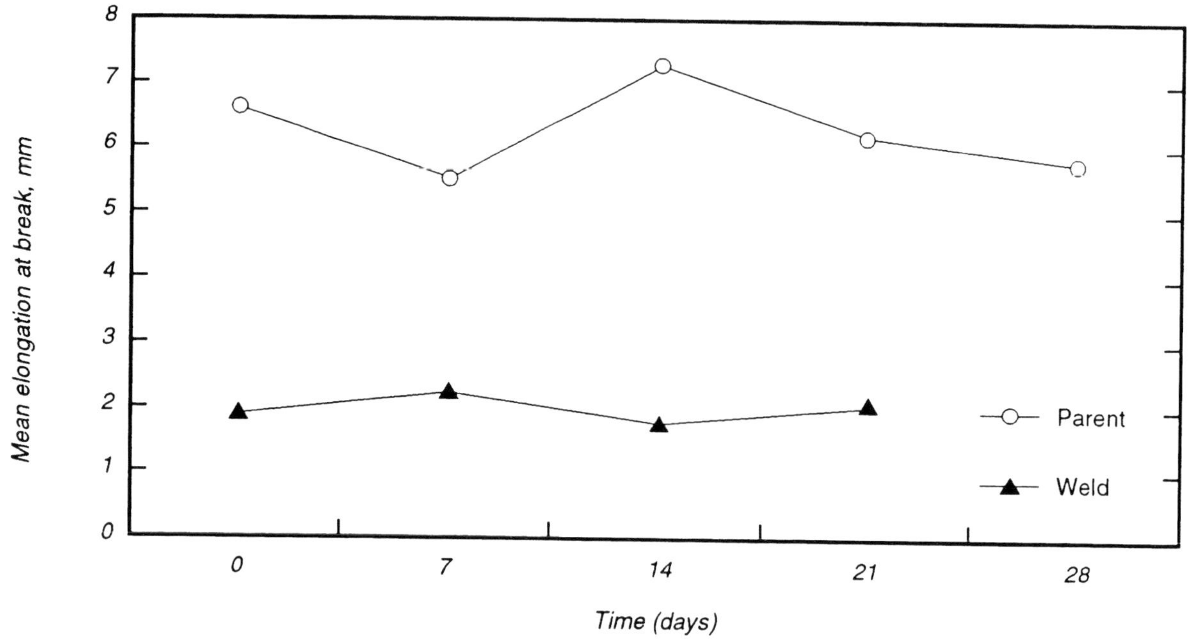

Fig. 14. Effect of immersion in boiling water on elongation at break of PEEK parent material and hot plate welds.

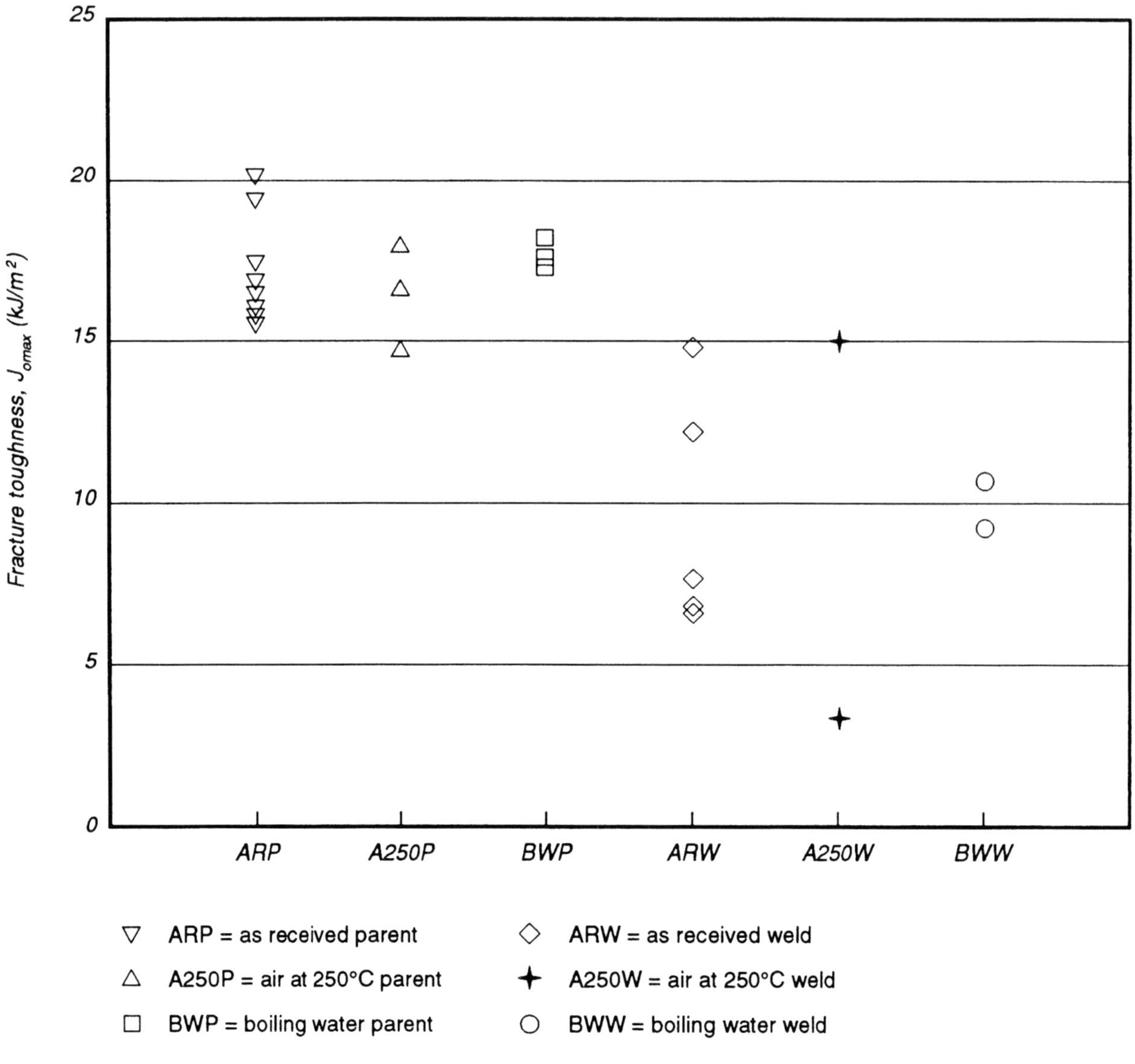

Fig. 15. The result of fracture toughness tests on PEEK parent materials and hot plate welds.

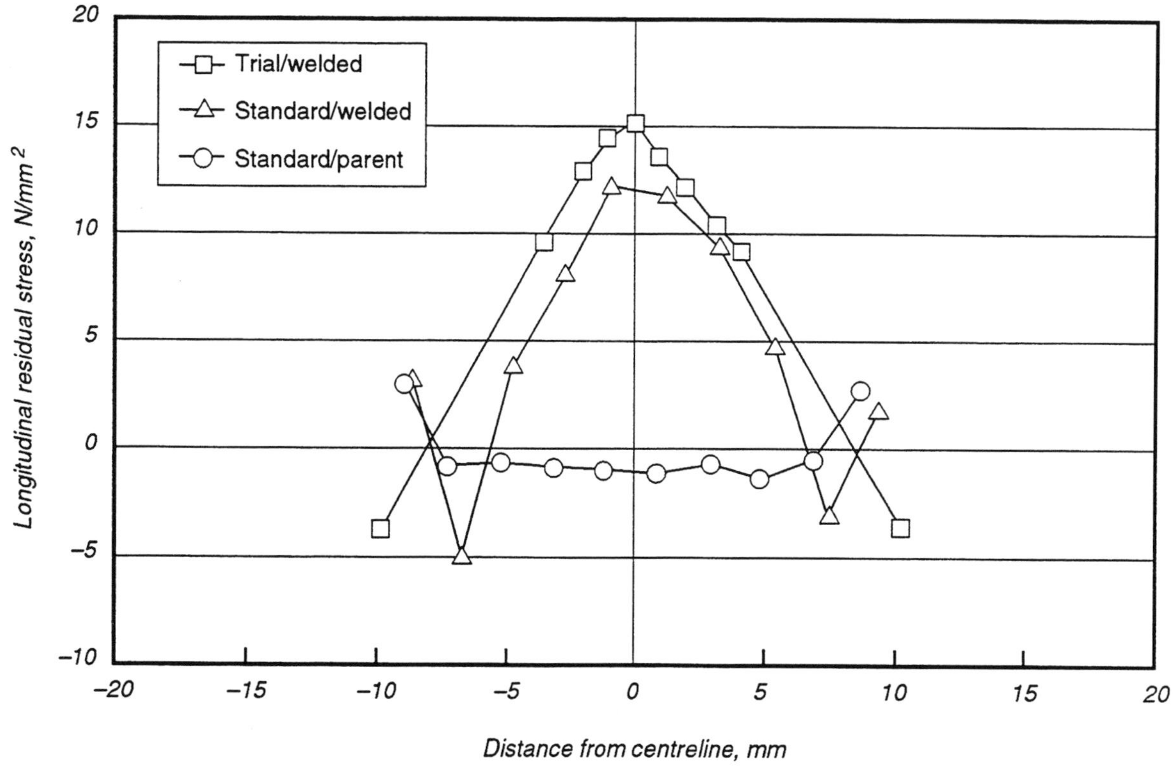

Fig. 16. Residual stresses in PEEK specimens: procedure development tests.

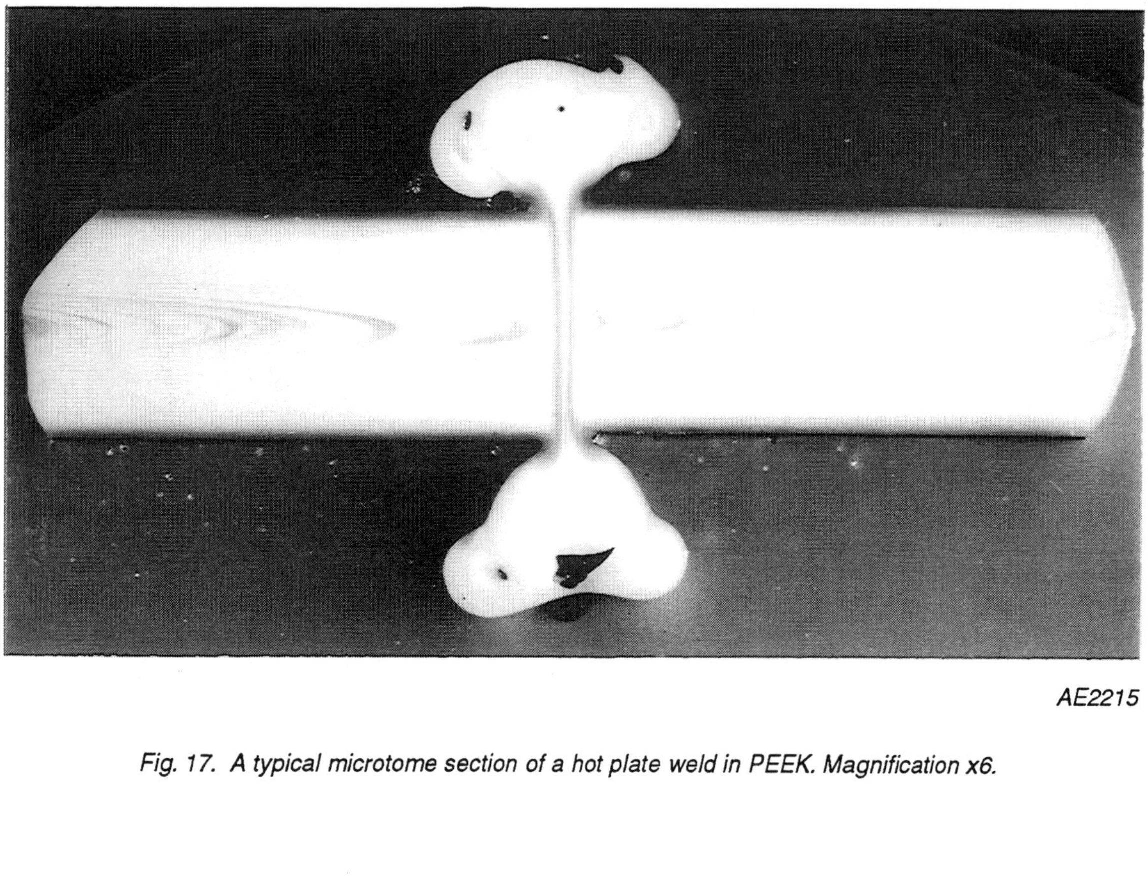

Fig. 17. A typical microtome section of a hot plate weld in PEEK. Magnification x6.

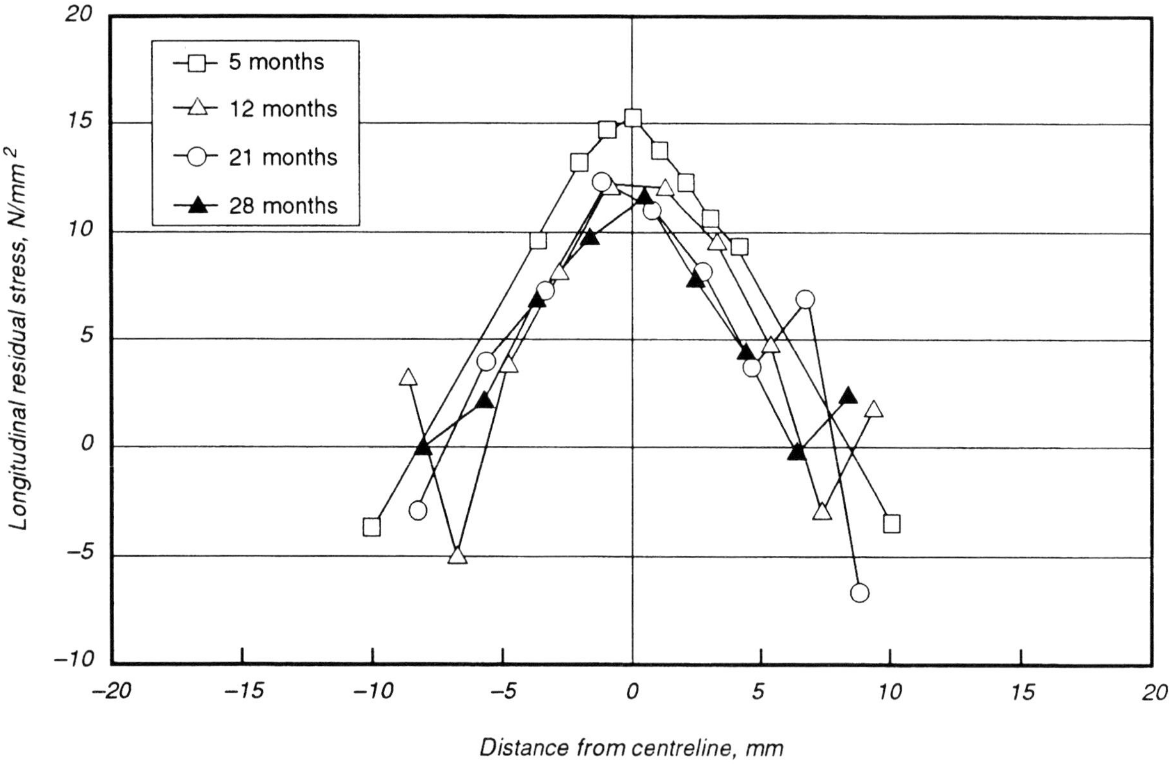

Fig. 18. Residual stresses in PEEK specimens 5 to 28 months after welding.

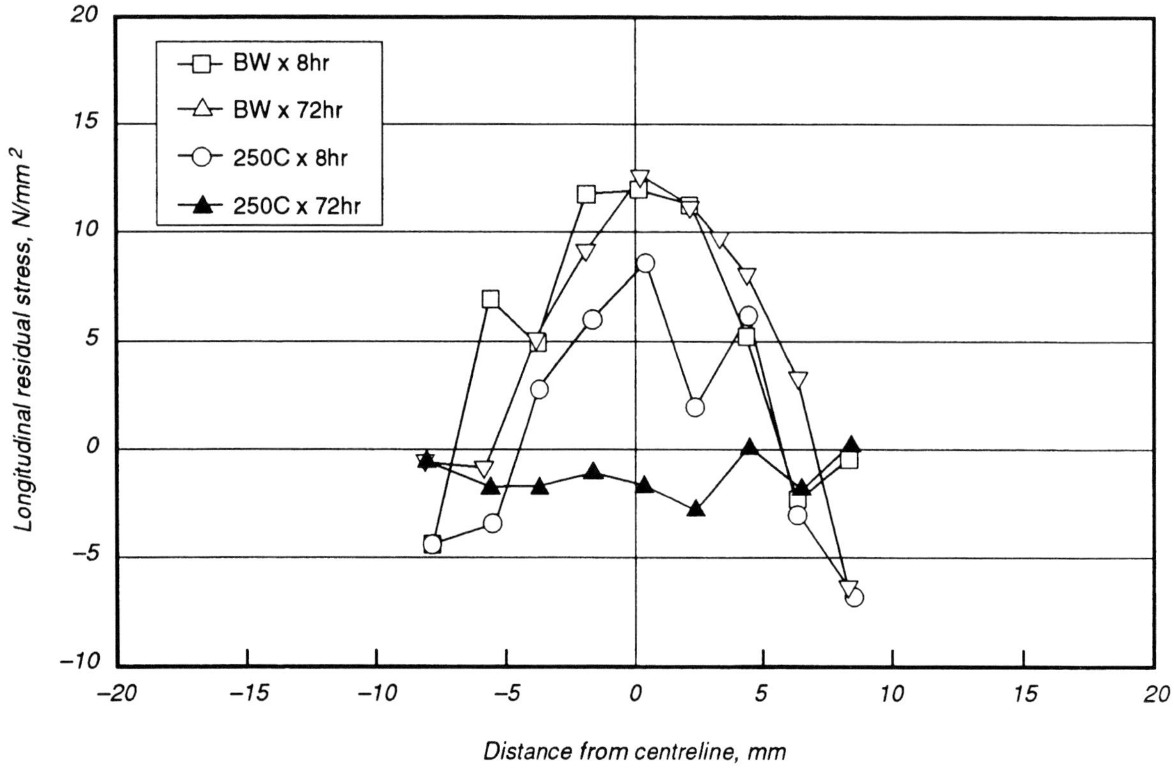

Fig. 19. Residual stresses in PEEK specimens after heat treatments.

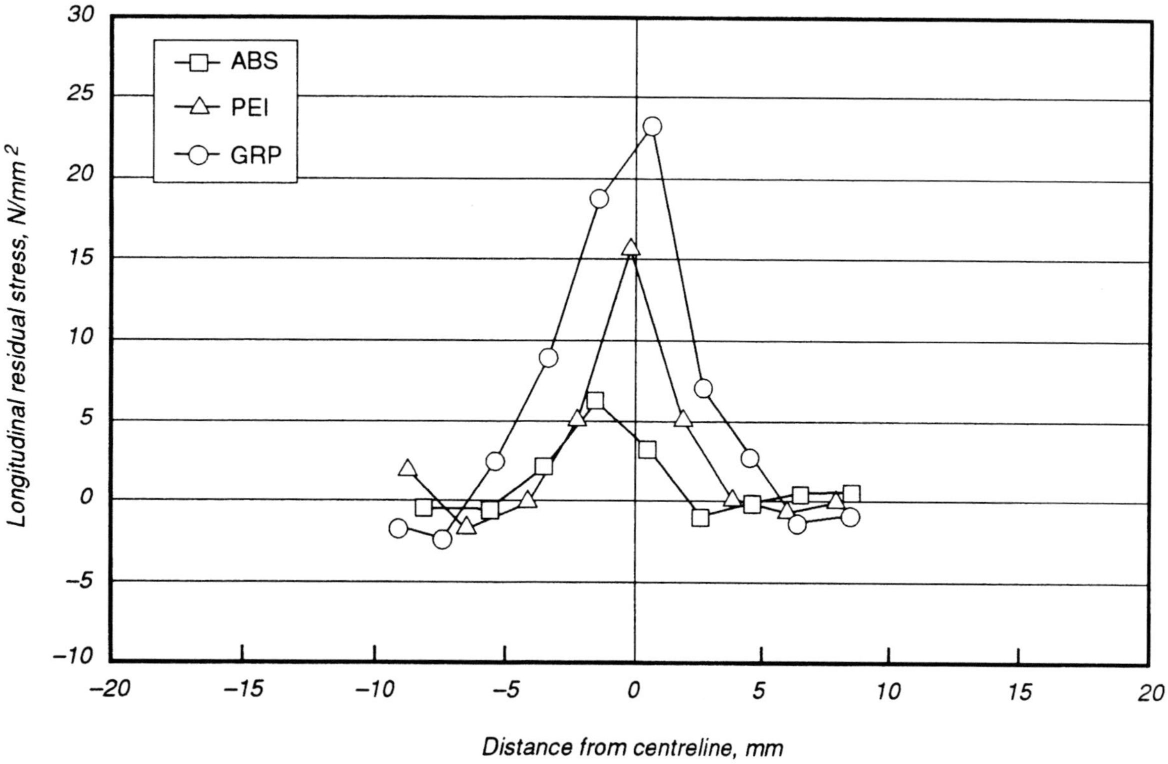

Fig. 20. Residual stresses in ABS, PEI and GR polyester specimens.

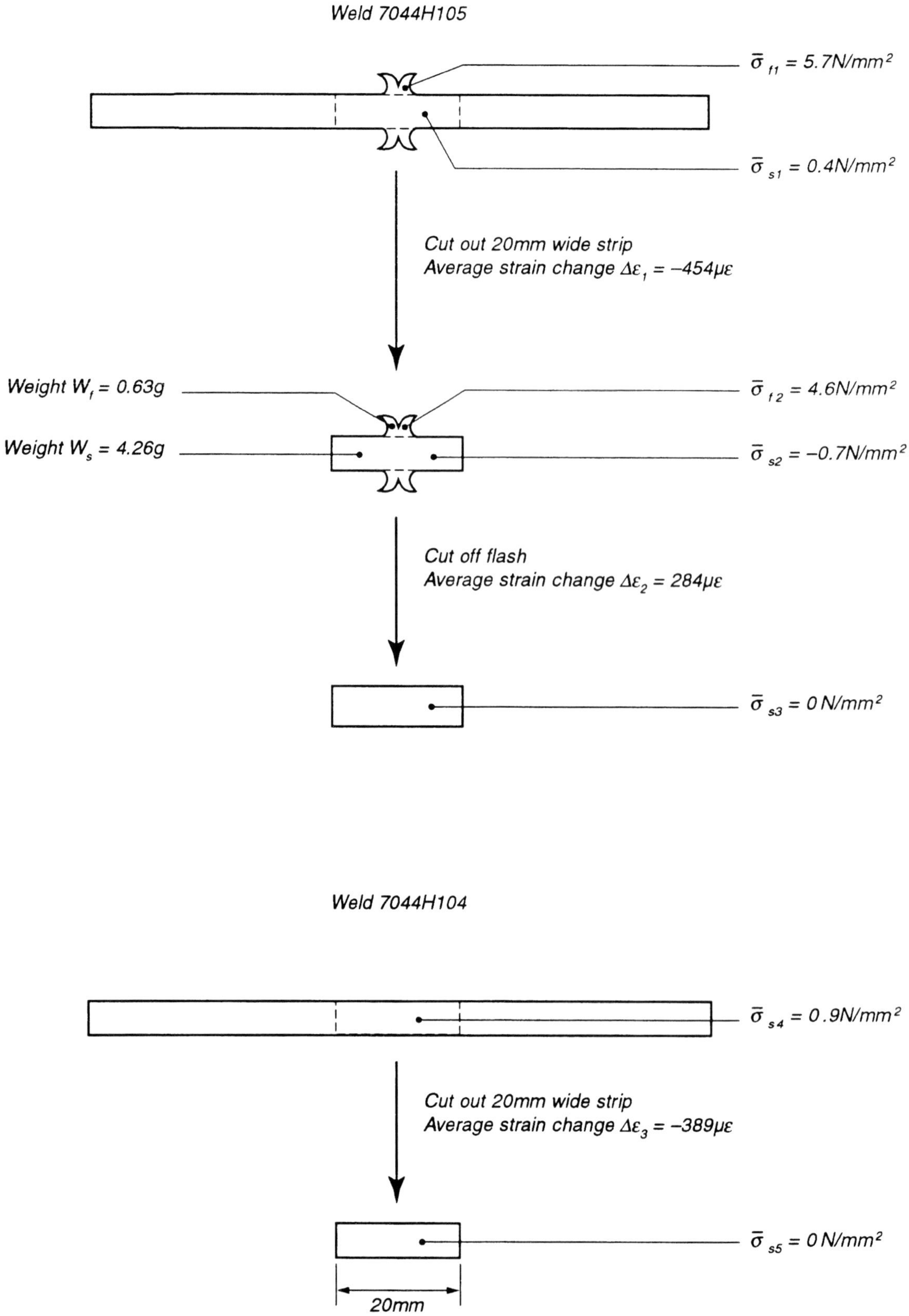

Fig. 21. Schematic of strain changes and average stresses in ABS specimens.

APPENDIX 1

BACKGROUND TO DEGRADATION OF PEEK

Although polymer degradation and stabilisation have been extensively studied in the past, interest and research in this field remain at a very high level [1–8]. There are a range of polymers which are fairly stable at normal atmospheric conditions. However, as the applications for polymeric materials widen, particularly in aerospace and electronic industries, the conditions under which they are expected to perform become more severe. A fundamental understanding of polymer degradation has led to new developments in polymers with high resistance to degradation and new stabilisers. The information obtained is also being used in many failure investigations to understand the mechanism of failure. Changes that occur in polymers during exposure to the environment are usually determined by comparison of the original properties with those measured at intervals during degradation. Any measurable property which is associated with degradation could be used to study polymer stability. These include determination of chemical composition, characterisation of chemical structure, measurement of mechanical properties, such as tensile strength at break, modulus at a given extension (eg. M100 at 100% elongation), elongation at break. In principal, the most appropriate accelerated ageing tests are based on conditions which are similar to those encountered during service but this is not always practicable. The proper selection of an experimental technique to determine the degree of ageing and to study degradation in polymers requires an understanding of the changes that occur in the chemical, physical and mechanical properties of the polymer upon ageing.

Polyetheretherketone (PEEK) is a high temperature resistant engineering thermoplastic. PEEK is a linear, crystallisable polymer with exceptional fracture toughness, delamination resistance, strength and rigidity as well as excellent chemical and radiation resistance. These properties combined with ease of fabrication, reprocessing capabilities and repairability mean that it has potential advantages over thermosets and some metals. A considerable number of papers have already been published on PEEK [18–28]. The following review will only be based on work reported on ageing and related area on PEEK.

Thermo–oxidative decomposition of PEEK in the presence and absence of copper was reported by Prime [18]. Thermogravimetric analysis (TGA) and mass spectrometry were used to investigate the degradation process, to identify gaseous products and examine the kinetics of the decomposition processes in PEEK. TGA showed that thermal stability of PEEK was diminished in air and dramatically reduced in the presence of air and copper. The differences were even more evident by isothermal TGA at 400°C. TGA–mass spectroscopy revealed the following decomposition products; phenol, diphenyl ether, benzophenone and benzoquinone. The presence of copper powder was observed to increase the degradation rate of PEEK at 375°C by 4 – 5 times. It was concluded that a small reduction in temperature or removal of oxygen from the processing environment should significantly reduce the decomposition rate in the presence of copper. The main reason for using copper in this work was to assess decomposition of PEEK when used in contact with copper as tooling material for processing.

Thermal degradation of PEEK was also investigated by Day et al [19]. Differential scanning calorimetry (DSC) was used to study the crystallisation kinetics and thermal characteristics of PEEK samples heated under a variety of conditions. This study showed that PEEK was sensitive to processing conditions i.e. melt temperature, melt time and environment. Changes in these conditions influenced the Tg, while the crystallisation process was also changed in different environments with oxidative reactions appearing to be the precursor of crosslinking reactions. It was concluded that, to minimise the effects of processing conditions on PEEK properties during fabrication, PEEK should be processed in a non–oxidative atmosphere preferably at temperatures below 400°C.

Thermogravimetry was also used by Hay and Kemmish [20] to study the thermal decomposition of PEEK. TGA was used to study the isothermal decomposition of PEEK and polylaryletherketone (PEK) and the kinetics of decomposition was established from the evolved volatiles. Both products and measured activation energies were consistent with a chain scission process occurring at ether and carbonyl linkages. The volatile decomposition products were analysed by mass spectroscopy and were found to contain various oligomers together with phenol and dibenzoturan. The volatile products were richer in hydrogen than the parent polymers, at the expense of a carbon–like insoluble residue. The degradation products were accounted for in terms of random chain scission and transfer reactions. It was concluded that degradation was initiated by random homolytic scission of either the ether or the carbonyl bonds in the polymer chains. The radicals produced abstract hydrogen from adjacent phenylene units, or terminate by combination to produce crosslinks.

Thermal stability and thermal decomposition of PEEK were also investigated by other workers [21 and 22]. Day and co–workers [21] studied the kinetic parameters of the thermal decomposition of PEEK in a nitrogen atmosphere using an isothermal weight–loss method. Several mathematical approaches were used to obtain the Arrhenius parameters based upon the size of different reaction models. It was found that the mechanism of degradation of PEEK was not a simple process, although the data could be interpreted on the basis of a unimolecular degradation.

The mechanism of thermal degradation of PEEK was also studied by Jonas and Legras [22]. A branching mechanism was observed, to occur in the usual melt processing conditions, which was much stronger in air than in vacuum. This degradation mechanism was correlated to a volatile emission observed with thermogravimetric analysis. The effect of the degradation on the PEEK crystallisation was also studied. A considerable decrease in crystallisation rate was observed, annotated with the decreased molecular mobility due to molecular weight increase, and a crystallinity decrease annotated with the structural defects introduced along the chains by the branching process.

Aromatic–based polymers such as PEEK have been finding application in many high–performance demanding engineering applications. These polymers are expected to be used in high radiation fields such as around fast–neutron breeder piles, fusion reactors, in space, or in medical applications in which exposure to γ–irradiation during sterilisation is required. Because of this, degradation of PEEK exposed to radiation has been studied by a number of workers [23–28]. The effect of radiation–induced oxidative irradiation effects (with γ–rays under oxygen pressure) on tensile properties of a number of aromatic–based polymers, including PEEK was studied by Susuga and Hagiwara [23]. The degradation resistance of PEEK to Co γ–radiation was found to be the highest compared to other aromatic–polymers (PES, polylarylester and polyarylamide) studied. The radiation stability of aromatic units under oxidative conditions was estimated from a comparison of the radiation resistance of aromatic units, resistance of the polymer themselves and their chemical structures. The following order was obtained: diphenyl ether, diphenyl ketone > aromatic amide > bisphenol A > diphenyl sulphone. The deterioration mechanism of PEEK under oxidative irradiation was studied by measuring dynamic viscoelastic properties. It was concluded that deterioration in mechanical properties under oxidative irradiation was brought about by chain scission only. In an earlier study Susagu and co-workers [24] evaluated the electron beam irradiation effects of ten types of polymers containing various aromatic rings on their tensile properties. The elongation at break was the most severely affected by the irradiation and it decreased with increasing dose. The order of radiation resistivity which was evaluated from the dose required for the elongation to change from 50% to 20% of the initial value was as follows:

Polymidies > PEEK > polyamide > polyetherimide > polyarylate > polysulphone > poly (phenylene oxide).

Space radiation effects on PEEK thin films and composites were also investigated by Funk and Sykees [25]. The influence of the films crystallinity on electron beam stability was evaluated using X–ray diffraction, DSC, Fourier–transform infrared (FTIR) spectroscopy and mechanical property tests. The mechanical properties of the composites material were evaluated after electron beam radiation followed by thermal cycling simulating 30 years in geosynchronous orbit. The result of this work indicated that both amorphous and semi–crystalline films of PEEK responded similarly to electron radiation. A significant electron interaction mechanism appeared to be a crosslinking process that resulted in a 50% decrease in film tensile elongation and slight changes in tensile strength and modulus. The threshold for major interaction was approximately 1 x 10^9 rads.

In a separate study, resistance of PEEK to photo–oxidation was evaluated [26]. The changes in surface chemistry of phenoxy resin and PEEK during photo–oxidation were monitored by X–ray photoelectron spectroscopy (XPS). Extensive oxygen uptake was observed for both materials and the data showed that oxidation of the phenyl rings had occurred. Comparison of phenoxy resin with its acetylated derivative indicated that the changes in surface chemistry of the former during the initial stages of exposure were not due to the hydroxyl group.

Environmental effect on the water absorption and mechanical properties of carbon fibre reinforced PEEK was studied by Ma and Yur [27]. It was shown that both humidity and temperature affected the structure and reduced the mechanical properties. However, the mechanical strength was still very high after long–term exposure. The retention of flexural strength was 95% and that of tensile strength was 70% after 1400 hours exposure.

In a similar study Buchman and Isayer [28] studied water absorption of some thermoplastics composites including PEEK laminates (APC–2/AS–4). The samples attained equilibrium after about 6 – 7 weeks. The absorption process was sharp at the beginning and asymptotic on approach to equilibrium. The APC–2/AS–4 absorbed about 0.2% water at equilibrium. This was due to the fact that PEEK is a semicrystalline polymer.

References

Please see main text for references.